Advances in Flotation of Copper, Lead and Zinc Minerals

Advances in Flotation of Copper, Lead and Zinc Minerals

Guest Editors

Qicheng Feng
Guang Han

MDPI

Basel • Beijing • Wuhan • Barcelona • Belgrade • Novi Sad • Cluj • Manchester

Guest Editors

Qicheng Feng
Faculty of Land Resource Engineering
Kunming University of Science and Technology
Kunming
China

Guang Han
Faculty of Land Resource Engineering
Kunming University of Science and Technology
Kunming
China

Editorial Office
MDPI AG
Grosspeteranlage 5
4052 Basel, Switzerland

This is a reprint of the Special Issue, published open access by the journal *Minerals* (ISSN 2075-163X), freely accessible at: https://www.mdpi.com/journal/minerals/special_issues/88EIKB8W6P.

For citation purposes, cite each article independently as indicated on the article page online and as indicated below:

Lastname, A.A.; Lastname, B.B. Article Title. *Journal Name* **Year**, *Volume Number*, Page Range.

ISBN 978-3-7258-4665-8 (Hbk)
ISBN 978-3-7258-4666-5 (PDF)
https://doi.org/10.3390/books978-3-7258-4666-5

Contents

About the Editors

Qicheng Feng

Qicheng Feng is a professor and doctoral supervisor at Kunming University of Science and Technology, where he also serves as the Deputy Director of the Yunnan Provincial Key Laboratory and the Academic Leader of the Ministry of Education's Engineering Research Center. His research focuses on the theory and technology of mineral separation, as well as the comprehensive utilization of complex and refractory mineral resources. He has been recognized as a Young Top Talent under the National High-Level Talents Support Program and as a Yunling Scholar under the Yunnan Provincial Talent Initiative. Prof. Feng serves as an editorial board member for several journals, including *International Journal of Minerals, Metallurgy and Materials*, and *Green and Smart Mining Engineering*. He has led over 20 major scientific projects funded by national and provincial programs, including the National Natural Science Foundation of China, National Key R&D Program sub-projects, Yunnan Provincial Science and Technology Programs, and industry–academia cooperation initiatives. His work has earned multiple prestigious awards, such as the First Prize of Yunnan Provincial Natural Science Award (2nd place), the First Prize of the Green Mine Science and Technology Award (1st place), and the Excellent Young Scientist Awards from both national- and industry-level associations. Prof. Feng has authored over 100 SCI-indexed papers as a first or corresponding author, including six ESI Hot Papers and eighteen ESI Highly Cited Papers. He holds 36 invention patents (China, USA, Russia, UK), of which 22 have been granted.

Guang Han

Guang Han is a Distinguished Professor and Master's Supervisor at Kunming University of Science and Technology. He has been recognized as a Young Talent under the Yunnan Provincial "Xingdian Talents" Program and is listed among the world's top 2% of scientists. His primary research interests include mineral separation theory and technology, and the comprehensive utilization of complex and refractory mineral resources. He has led several scientific projects supported by the National Natural Science Foundation of China, the Yunnan Basic Research Programs, and the State Key Laboratory Youth Science Fund. Dr. Han has published over 30 SCI-indexed papers as a first or corresponding author, including one ESI Hot Paper and five ESI Highly Cited Papers. He has been granted 19 invention patents, and has authored one academic monograph as the lead editor. His work has been recognized with the First Prize of the Green Mine Science and Technology Award (4th place), the Yunnan Provincial Excellent Doctoral Dissertation Award, and the Excellent Doctoral Dissertation Award of the Nonferrous Metals Society of China. He serves on the editorial boards or as a youth editor for journals such as *Gold*, *International Journal of Mining Science and Technology*, *International Journal of Minerals, Metallurgy and Materials*, *Journal of Central South University*, *Green and Smart Mining Engineering*, *cMat*, and *Journal of Central South University (Natural Science Edition)*.

Preface

This Special Issue reprint, titled "Advances in Flotation of Copper, Lead and Zinc Minerals," collates a collection of ten high-quality contributions that reflect the latest advances in froth flotation for non-ferrous metal ores. The aim of this Special Issue is to showcase innovative research that addresses the increasing complexity of ore characteristics and the urgent demand for more sustainable and efficient separation technologies. The topics span fundamental surface chemistry, reagent development, flotation hydrodynamics, and practical flowsheet optimization, offering valuable insights for both academic researchers and industrial practitioners.

We were motivated to compile this Special Issue due to the pressing need for scientific and technological innovation in the field of base metal beneficiation, especially in light of the declining ore grades and environmental considerations. We believe this reprint will serve as a valuable resource for flotation researchers, process engineers, and graduate students seeking a deeper understanding of and solutions for Cu-Pb-Zn processing.

We sincerely thank all of the contributing authors for their excellent work and the reviewers for their constructive feedback. We also gratefully acknowledge the *Minerals* editorial team for their support throughout the process.

Qicheng Feng and Guang Han
Guest Editors

minerals

MDPI

Editorial

Editorial for Special Issue "Advances in Flotation of Copper, Lead and Zinc Minerals"

Qicheng Feng * and Guang Han

State Key Laboratory of Complex Nonferrous Metal Resources Clean Utilization, Faculty of Land Resource Engineering, Kunming University of Science and Technology, Kunming 650093, China; ghkmust@126.com
* Correspondence: fqckmust@163.com

Copper, lead, and zinc metals are fundamental pillars of modern industry, yet their extraction faces increasing hurdles. Froth flotation, the dominant concentration technology, relies on exploiting subtle differences in surface properties. However, the depletion of high-grade ores compels the industry to tackle more complex, lower-grade, and often oxidized or finely disseminated mineral resources. Concurrently, the drive for sustainability necessitates more selective and efficient flotation processes with minimized environmental footprints. This convergence of challenges demands continuous innovation in flotation science and technology specific to copper, lead, and zinc minerals. This Special Issue, "Advances in Flotation of Copper, Lead, and Zinc Minerals", aims to consolidate recent progress, presenting research that spans from fundamental mechanisms to practical process optimization for these vital non-ferrous metals.

The contributions gathered herein reflect the multifaceted nature of current research in Cu-Pb-Zn flotation. A significant focus remains on the development and mechanistic understanding of novel flotation reagents designed to enhance selectivity in difficult separations. Addressing the classic challenge of separating smithsonite from calcite, Wang et al. [1] demonstrate the efficacy of the chelating agent cupferron, meticulously elucidating its selective adsorption mechanism through comprehensive surface analyses (FTIR, zeta potential, and XPS). The broader potential of chelating agents, not only as collectors but also as activators or depressants in zinc oxide systems, is systematically reviewed by Song et al. [2], offering a valuable guide for future reagent design.

Moving beyond single reagents, the synergistic potential of collector mixtures is explored. Li et al. [3] present an optimized open-circuit strategy for zinc oxide ore beneficiation using a ternary collector system (DDA + NaIX + ADD). Their work underscores how combined collectors can enhance hydrophobicity and improve flotation kinetics compared to single or binary systems, supported by detailed adsorption studies. Similarly, Wang et al. [4], while focusing on the reverse flotation of hematite, investigate the interaction of mixed anionic/cationic collectors (DTAC/Tall oil) on quartz. Understanding such interactions on quartz, a ubiquitous gangue mineral often associated with Cu-Pb-Zn ores, provides transferable insights into managing gangue behavior in base metal flotation circuits.

The selective depression of undesired minerals is equally crucial. Two studies by Zeng et al. [5,6] introduce novel depressants for the selective separation of galena from chalcopyrite. The first employs PMA–EDTC, a functionalized macromolecular depressant, demonstrating its selective chemisorption on galena via its –N–(C=S)–S– group through

FTIR, XPS, and DFT studies. The second investigates thioureidoacetic acid (TA), a small-molecule depressant with dual functional groups (C=S, C=O), again using a multi-technique approach including DFT with a COHP analysis to reveal strong, dual-point chemisorption (Pb–S, Pb–O bonds) on galena, establishing a compelling case for molecular orbital engineering in depressant design.

Fundamental parameters governing flotation response are also critically examined. The pervasive influence of pulp chemistry, specifically the role of various metal ions (Cu^{2+}, Pb^{2+}, Fe^{2+}/Fe^{3+}, Ca^{2+}, Mg^{2+}, etc.), is comprehensively reviewed by Wei et al. [7]. They detail the diverse sources of these ions and their complex activation or depression mechanisms on different Cu, Pb, and Zn minerals, highlighting the necessity of considering ionic interactions in process control. Process conditions like pH are universally critical; Wang et al. [8] investigate the specific inhibiting mechanism of high pH on molybdenite flotation, using experimental and DFT methods to differentiate the response of its distinct basal and edge surfaces (MS001 vs. MS100). While focused on molybdenite, the study provides fundamental insights into how pH and consequent surface hydroxylation can differentially affect anisotropic mineral surfaces—a concept relevant to layered or variably terminated base metal sulfides as well.

Furthermore, the physical aspects of flotation hydrodynamics, particularly bubble size, are essential for optimizing particle capture, especially with emerging coarse particle flotation technologies. Gahona et al. [9] characterize bubble size distribution within a HydroFloat® fluidized-bed cell, comparing tap water and seawater systems with different frothers. Their findings offer valuable data on bubble behavior in this specific equipment under varying water qualities, directly relevant to applying such technologies efficiently in Cu-Pb-Zn circuits, particularly those operating in regions facing water scarcity or processing coarser feeds.

Finally, bridging fundamental understanding with practical application, Yu et al. [10] present a detailed study integrating process mineralogy (XRF, XRD, and EPMA) with flotation optimization for a challenging low-grade oxidized lead–zinc ore from the Lanping mine. This work exemplifies the crucial link between thoroughly characterizing the ore's specific mineralogical features—composition, liberation, and intergrowths—and developing a tailored, effective flotation flowsheet, ultimately employing mixed depressants and specialized collectors to achieve separation.

In summary, this Special Issue encapsulates key advancements in the flotation of copper, lead, and zinc minerals. The collective works span the molecular design and mechanistic elucidation of novel, selective collectors and depressants; comprehensive reviews of fundamental factors like chelating agents and metal ion chemistry; investigations into process hydrodynamics; and ore-specific treatment strategies guided by process mineralogy. The sophisticated integration of advanced analytical techniques (XPS, FTIR, ToF-SIMS, and EPMA) and computational methods (DFT and MD) is evident throughout, providing deeper insights into complex interfacial phenomena. While significant progress has been demonstrated, the translation to robust industrial practice—considering economic factors, reagent synthesis scalability, environmental profiles, and adaptation to diverse ore types and process waters—remains the critical next step. We trust this collection will serve as a valuable resource and stimulus for continued innovation in the sustainable processing of copper, lead, and zinc resources.

Conflicts of Interest: The authors declare no conflict of interest.

References

1. Wang, Q.; Sun, L.; Cao, Y.; Wang, X.; Qiao, Y.; Liu, G.; Xiang, M.; Sun, W. Flotation Separation of Smithsonite from Calcite Using Cupferron as a Collector. *Minerals* **2023**, *13*, 992. [CrossRef]
2. Song, Z.; Wen, S.; Han, G.; Feng, Q. Recent Progress on Chelating Reagents in Flotation of Zinc Oxide Ores: A Review. *Minerals* **2023**, *13*, 1278. [CrossRef]
3. Li, Z.; Feng, Q.; Zhang, Q.; Wen, S. Open-Circuit Technology of Zinc Oxide Ore Flotation with Ternary Collector and Its Adsorption Characteristics on Smithsonite Surface. *Minerals* **2024**, *14*, 902. [CrossRef]
4. Wang, J.; Lu, D.; Han, X. Application and Mechanism of Mixed Anionic/Cationic Collectors on Reverse Flotation of Hematite. *Minerals* **2024**, *14*, 550. [CrossRef]
5. Zeng, H.; Zhu, Y.; Sun, C.; Zhao, Z.; Wu, G.; Liu, C.; Lu, T.; Zhang, X. Utilization of PMA-EDTC as a Novel Macromolecular Depressant for Galena in the Flotation Separation of Chalcopyrite. *Minerals* **2023**, *13*, 1478. [CrossRef]
6. Zeng, H.; Liu, C.; Lu, T.; Gao, Z.; Zhu, Y.; Sun, C.; Zhao, Z.; Wu, G.; Li, R.; Hu, J. Investigation of a Novel Depressant for Flotation Separation of Chalcopyrite and Galena: Experiments and Adsorption Mechanisms. *Minerals* **2025**, *15*, 454. [CrossRef]
7. Wei, B.; Mao, Y.; Wang, L.; Shen, P.; Wang, H.; Liu, D. Sources, Performance and Mechanisms of Metal Ions in the Flotation Process of Copper, Lead, and Zinc Ores: A Review. *Minerals* **2024**, *14*, 1105. [CrossRef]
8. Wang, E.; Wan, H.; Qu, J.; Yi, P.; Bu, X. Inhibiting Mechanism of High pH on Molybdenite Flotation. *Exp. DFT Study. Miner.* **2024**, *14*, 663.
9. Gahona, G.; Cisternas, L.A.; Araya-Gómez, N.; Lucay, F.A.; Gálvez, E.D.; Lopéz-Valdivieso, A.; Valdes, F. Bubble Size Characterization in the HydroFloat® Fluidized-Bed Flotation Cell Using Tap Water and Seawater. *Minerals* **2024**, *14*, 813. [CrossRef]
10. Yu, A.; Ding, Z.; Yuan, J.; Feng, Q.; Wen, S.; Bai, S. Process Mineralogy Characteristics and Flotation Optimization of a Low-Grade Oxidized Lead and Zinc Ore from Lanping Mine. *Minerals* **2023**, *13*, 1167. [CrossRef]

minerals

MDPI

Review

Recent Progress on Chelating Reagents in Flotation of Zinc Oxide Ores: A Review

Zhengyong Song [1,2], Shuming Wen [1,2], Guang Han [1,2,*] and Qicheng Feng [1,2,*]

1 State Key Laboratory of Complex Nonferrous Metal Resources Clean Utilization, Faculty of Land Resource Engineering, Kunming University of Science and Technology, Kunming 650093, China; zysongkust@126.com (Z.S.); shmwen@126.com (S.W.)
2 Yunnan Key Laboratory of Green Separation and Enrichment of Strategic Mineral Resources, Kunming University of Science and Technology, Kunming 650093, China
* Correspondence: ghkmust@126.com (G.H.); fqckmust@163.com (Q.F.)

Abstract: Zinc oxide minerals (primarily smithsonite and hemimorphite) are important sources of Zn. Flotation is the most widely used method of enriching zinc oxide minerals for mineral processing. Chelating reagents have received extensive attention for the flotation of zinc oxide ores because of their high selectivity and stability. This paper systematically summarizes the selective separation performance and mechanisms of various chelating reagents as collectors, activators, and depressants in the flotation of zinc oxide ores. The types and mechanisms of chelating reagents in the flotation of zinc oxide ores are discussed, providing new ideas for the development of efficient flotation reagents for zinc oxide ores.

Keywords: zinc oxide ores; flotation; chelating reagents; selective separation

1. Introduction

Zinc is widely distributed in nature. It possesses several physical and chemical characteristics, including good electrical and thermal conductivities, corrosion resistance, and ductility. These properties make it suitable for various applications, such as galvanizing processes and alloy production, and as a chemical additive in fields such as medicine, rubber, and paint [1–4]. In recent decades, the global consumption of Zn has significantly increased, making it the fourth most consumed metal after iron, aluminum, and copper [5]. Zn resources are primarily found in sulfide deposits, with sphalerite being the main Zn-bearing mineral. Therefore, the main raw materials used in industrial production are zinc sulfide minerals. However, the continuous consumption of limited zinc sulfide resources has made it difficult to meet the growing demand for Zn. Consequently, the efficient development and utilization of zinc oxide ores have become crucial for supplementing the supply of Zn metal [6].

Several types of zinc-bearing minerals exist in zinc oxide ores, including smithsonite ($ZnCO_3$), willemite (Zn_2SiO_4), hemimorphite ($Zn_4Si_2O_7(OH)_2 \cdot H_2O$), hydrozincite ($3Zn(OH)_2 \cdot 2ZnCO_3$), and zincite ($ZnO$) [7–9]. The most prevalent and economically valuable among them are smithsonite and hemimorphite. In nature, zinc oxide minerals do not exist alone. The minerals associated with zinc oxide minerals mainly include copper, lead, and iron oxide minerals, such as malachite, cerussite, and limonite. The main gangue minerals are calcite, dolomite, quartz, goethite, and clay minerals (kaolinite, chlorite, mica, etc.). Because of the complex relationship between the ore structure and co-association in zinc oxide ores, the surface properties of the target mineral and gangue are similar (crystal structure, solubility, and hydrophobicity), and they have a fine particle size, severe mud phenomenon, and high soluble salt content, leading to difficulties in Zn recycling [5,7]. Flotation is the main method used to treat zinc oxide ores in the field of mineral processing [10], and the treatment of zinc oxide ores usually includes the following methods:

(1) Sulfidization–xanthate flotation: Sulfidization–xanthate flotation is an effective method for treating zinc oxide ores. It uses a sulfiding agent to pre-sulfurize the surfaces of oxide minerals, covering them with a hydrophobic sulfide film, and then using xanthate as a collector for flotation [11]. However, when sulfidization–xanthate flotation is used to directly separate zinc oxide ores, the flotation index is often not ideal and is difficult to directly apply in industrial production. To improve the flotation efficiency and mineral recovery of sulfidization–xanthate flotation, it is usually necessary to heat and add metal ions for activation [9,12]. Complex procedures, high reagent dosages, and unreliable industrial production indices may result from the additional processing steps [13].

(2) Sulfidization–amine flotation: Sulfidization–amine flotation is the most commonly used flotation technology for zinc oxide ores in industry; it first sulfidizes zinc oxide minerals with sulfidizing reagents (Na_2S, NaHS, etc.) and then uses cationic collectors (such as dodecylamine and octadecylamine) for flotation [14,15]. When using this method, it is necessary to strictly control the dosage of sulfidizing reagents and the pH of the flotation pulp [16]. In addition, there are problems such as high reagent consumption, difficulty in defoaming, sensitivity of amine collectors to pulp, and soluble salts during the flotation process.

(3) Direct flotation via fatty acids: Fatty acids can be used for the direct flotation of zinc oxide minerals. This technology performs well in the flotation of zinc oxide ores containing silicate or muddy gangue, but has poor selectivity for zinc oxide ores containing carbonate gangue [17,18].

(4) Thiol collector flotation: Non-aromatic compounds (thiols) containing thiol functional groups (−SH) were used as collectors for zinc oxide mineral flotation, and ideal flotation results were obtained [19]. Hosseini et al. [9] pointed out that after thiol was adsorbed on the surface of zinc oxide minerals, the S–H bond in thiol broke. The broken H reacted with −OH on the surfaces of the minerals to form H_2O, and the broken S formed a S–Zn bond with Zn on the surfaces of the minerals. The S atom is connected to the hydrophobic group. Therefore, the formation of the S–Zn species enhances the hydrophobicity of the mineral surface and realizes the collection of zinc oxide minerals by the thiol collector. However, in the actual flotation process, when thiol is used as a collector to float smithsonite, the flotation recovery obtained is low. In addition, thiol also has the disadvantages of easy volatilization, an unpleasant smell, and a large dosage consumption. The industrial production index is not stable enough, so it is less used in industry.

(5) Chelating reagent flotation: Chelating reagent flotation uses specific chelating reagents as collectors or regulators to achieve efficient mineral separation. Owing to their exceptional selectivity and stability, chelating reagents have been extensively studied as collectors or regulators for the flotation of zinc oxide ores [20,21].

Chelating reagents can interact with metal ions to form stable complexes with ring structures. Chelating reagents generally contain two or more electron-donating groups that combine with metal ions to form cyclic structures called metal chelates [22–25]. The active atoms in the chelating reagents provide electron pairs that interact with the metal ions. These active atoms are often non-metallic elements, mainly O, N, S, and P atoms [22,24]. Electron-donating groups can be classified into acidic and basic groups. Acidic groups can lose protons and coordinate with metal atoms. Common acidic groups include the −COOH (carboxylic), $-SO_3H$ (sulfonic), $-PO(OH)_2$ (phosphoric), −OH (enolic and phenolic), =N–OH (oxime), and −SH (thiol) groups. Basic groups contain atoms carrying lone pairs of electrons that can interact with metal cations. Common basic groups include −NH− (imino), $-NH_2$ (amino), −N= (tertiary acyclic or heterocyclic nitrogen), =O (carbonyl), −O− (ether or ester), =N–OH (oxime), −OH (aliphatic alcohol), etc. [24,26].

Recently, the development of efficient flotation reagents for oxide minerals has become prevalent in mineral processing. Owing to the greater stability of metal chelates, chelating reagents have been widely investigated as flotation reagents in mineral processing and are considered to have excellent collection and selective properties with enormous potential for applications. For example, hydroxamic acid has been widely applied in the flotation of oxide minerals such as scheelite, wolframite, and cassiterite [27–29]. Chelating reagents have also been extensively investigated for the flotation of zinc oxide minerals and have achieved significant progress and advancements. For example, hydroxamic acid, phosphate, amide, 8−hydroxyquinoline, and heterocyclic collectors have been widely investigated in the flotation of zinc oxide minerals, and some chelating reagents containing oxime groups, hydroxyl groups, carboxyl groups, and phosphate groups have also been investigated as regulators in the flotation of zinc oxide minerals. However, a systematic introduction to the application and mechanisms of action of chelating reagents in the flotation of zinc oxide minerals has rarely been reported. This article summarizes the selective separation performance and action mechanisms of various chelating reagents in the flotation of typical zinc oxide and gangue minerals, providing new ideas and a theoretical basis for the study and application of chelating reagents in the flotation of zinc oxide ores.

2. Crystal Structures and Surface Properties of Typical Zinc Oxide Minerals and Gangue Minerals

2.1. Crystal Structures and Surface Properties of Typical Zinc Oxide Minerals

Smithsonite is a typical island carbonate mineral with a trigonal crystal structure. As illustrated in Figure 1a, Zn, in a single smithsonite unit cell, coordinates to O in the surrounding six CO_3^{2-} groups to form six-coordinate bonds, whereas C coordinates to O to form three-coordinate bonds. Furthermore, O coordinates to two surrounding Zn atoms and one C to form a three-coordinate structure. The strength of the Zn–O bond in the smithsonite crystal is weaker than that of the C–O bonds in the carbonate groups. As a result, the Zn–O bond is more prone to fracture during crushing and grinding, leading to the exposure of fresh Zn sites on the mineral surface [30,31]. Smithsonite is a semi-soluble mineral with a solubility product constant of 1.46×10^{-10}. Due to the dissolution and hydration of lattice ions, there is a large number of unavoidable ions in the solution, and the dominant components in the solution include $ZnOH^+$, $Zn(OH)_2$, $Zn(OH)_3^-$, and $Zn(OH)_4^{2-}$ [32–34]. The hydrolysis products can be adsorbed to the mineral surface, thus making the flotation separation of smithsonite and gangue minerals more difficult. Moreover, the smithsonite surface is partially hydroxylated to form zinc hydroxides in alkaline conditions, which hinders the interaction between flotation reagents and active sites on the surfaces of the minerals.

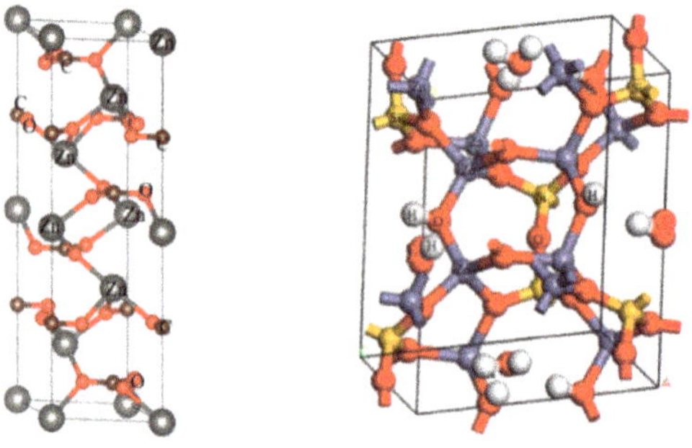

(a) (b)

Figure 1. (**a**) Crystal structure of smithsonite [31]; (**b**) cell model of hemimorphite [35].

Hemimorphite is a complex silicate mineral whose structure is shown in Figure 1b. In the crystal structure of hemimorphite, Zn atoms form new tetrahedra with –OH and three O atoms belonging to the Si–O tetrahedron, resulting in low Zn activity in the lattice [35]. The bonding strength between the Si–O bonds being stronger than that between the Zn–O bonds leads to the breaking of the Zn–O bonds during the crushing and grinding process of hemimorphite. Consequently, the SiO_4^{2-} and Zn^{2+} chemical bonds are often exposed on the surfaces of minerals. These SiO_4^{2-} components often shield the Zn component on the surface of the hemimorphite, reducing the reactivity of the surface with the flotation reagents and making flotation difficult.

2.2. Crystal Structure and Surface Properties of Common Gangue Minerals

Calcite is the most common gangue mineral in zinc oxide ores. It belongs to the trigonal crystal system with the space group symbol R3C and has a closely packed structure. As shown in Figure 2, Ca^{2+} ions in the calcite crystal lattice combine with O atoms in the surrounding six $CO_3{}^{2-}$ ions to form an irregular octahedral coordination with a Ca–O bond length of 2.36 Å. The C atom is coordinated with the O atoms to form a structure with a C–O bond length of 1.28 Å [36]. The cleavage planes of calcite are along three groups, namely, the {101}, {211}, and {102} surfaces of the rhombohedral plane, which correspond to the {104}, {018}, and {214} surfaces of the hexagonal cell [37]. The crystal structures and surface properties of the calcite and smithsonite are similar. In pulp solutions, the surfaces of calcite and smithsonite can be hydroxylated and dissolved, affecting the flotation separation of minerals through the release and migration of calcium ions and the heterogeneous coagulation of minerals.

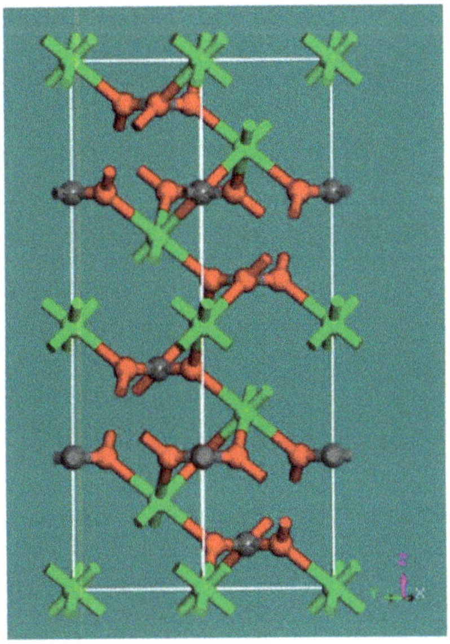

Figure 2. Unit cell structure of calcite (Ca = light green, O = red, C = grey) [36].

Quartz is another common gangue mineral in zinc oxide ores. Quartz surfaces possess strong electronegativity, a low isoelectric point, and high hydrophilicity. Quartz does not have a fixed cleavage surface, and researchers typically choose to analyze the structure and surface properties of quartz on the (101), (100), and (001) surfaces. Quartz is composed of Si–O tetrahedrons that form a frame-like crystal structure. Each Si atom is bonded to four O atoms through four single bonds, forming a Si–O tetrahedron [38]. The Si–O bond energy was high, which caused a large number of Si–O bonds in the quartz crystal to break during the crushing and grinding processes, leading to the exposure of Si^{4+} and O^{2-} ions on the surface. These ions can adsorb OH^- and H^+ in aqueous solutions, resulting in the formation of hydroxylated quartz surfaces (Figure 3) [39–41]. During the flotation process, the Si–O functional group on the quartz surface can react with the hydroxyl complexes of metal ions, causing the metal ions to adsorb onto the quartz surface and form the action sites of the flotation reagents, thereby activating flotation [38,42,43].

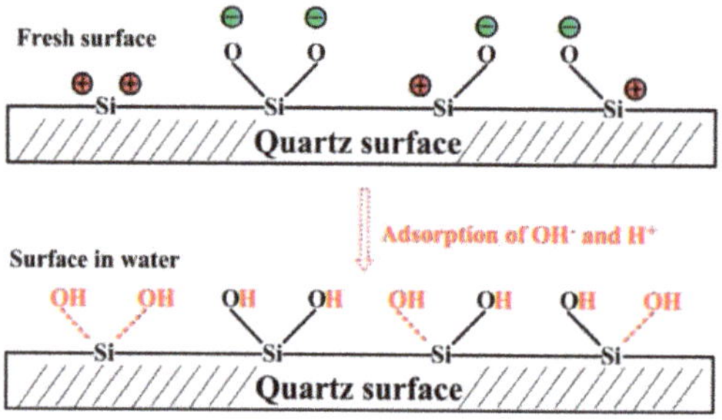

Figure 3. Quartz surface model [41].

3. Chelating Collectors

Based on the different types of electron donor atoms in the chelating collectors, various types of chelating collectors can be distinguished for the flotation of zinc oxide ores. These include O–O, N–O, S–S, and S–N chelating collectors (Table 1).

Table 1. Common chelating collectors for zinc oxide minerals.

Type	Name	Mineral	Optimum pH Range	Molecular Structure	Refs.
O–O	Salicylhydroxamic acid	Smithsonite	5–7.5	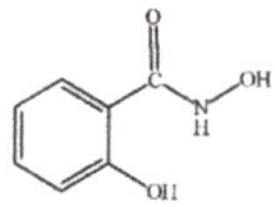	[44]
	Octyl hydroxamic acid	Smithsonite	7.5–8.5	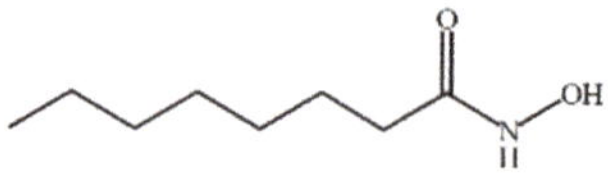	[45]
	Salicylaldoxime	Smithsonite	7.5–8.5	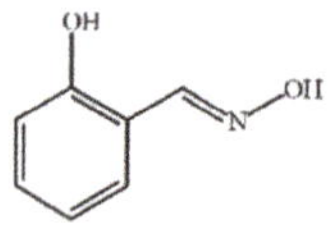	[46]
	Benzohydroxamic acid	Smithsonite	7–9	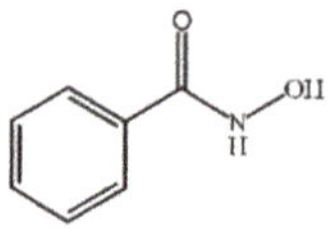	[46,47]
	Phenylpropenyl hydroxamic acid	Smithsonite	7	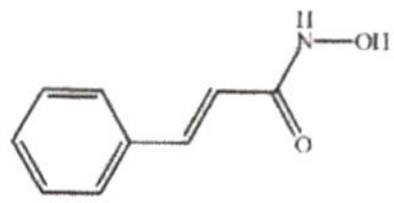	[48]
	Cupferron	Smithsonite	7.5–9	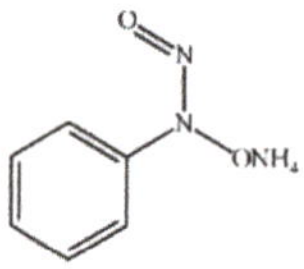	[49]
	Dodecylphosphate	Smithsonite	8–10	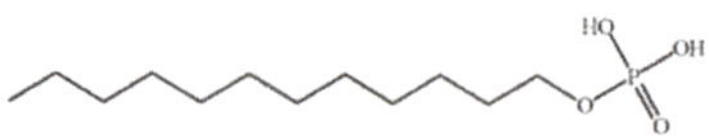	[50]
	Di(2-ethylhexyl)phosphoric acid	Hemimorphite	6–10	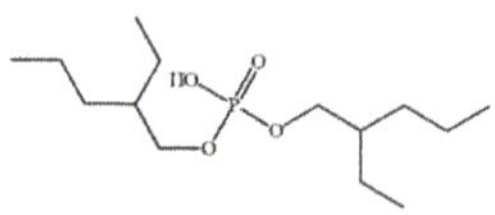	[51]

Table 1. *Cont.*

Type	Name	Mineral	Optimum pH Range	Molecular Structure	Refs.
N–O	N-lauroylsarcosinate	Hemimorphite	7	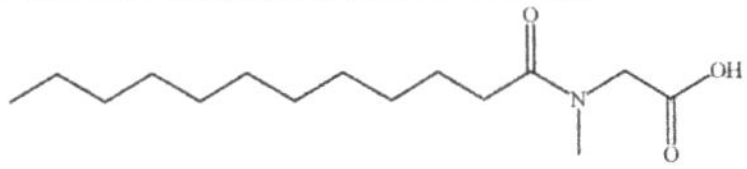	[52–54]
	8-Hydroxyquinoline	Smithsonite	7	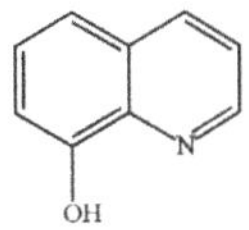	[55–57]
S–S	Mercaptobenzothiazole	Smithsonite	<7	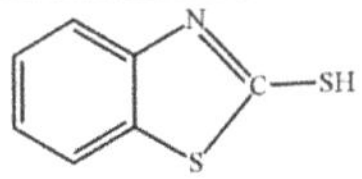	[58]
S–N	Aminothiophenol	Smithsonite	–	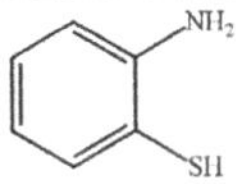	[59]

3.1. O–O Chelating Collectors

Hydroxamic acid is widely used as a chelating collector in the flotation of zinc oxide ores. The −C(O)NHOH group in its molecular structure possesses high coordination activity, providing strong binding to metal ions on mineral surfaces and forming stable chelates, which enable stable adsorption onto mineral surfaces and enhance their hydrophobicity [48,60].

Zhao et al. [44] discovered that using salicylhydroxamic acid as a collector in the flotation of smithsonite resulted in a recovery of over 90% within a pH range of 5–7.5, demonstrating the strong collection ability of salicylhydroxamic acid for smithsonite. The zeta potential and infrared spectral analyses confirmed that salicylhydroxamic acid interacted with the surface of smithsonite through chemical adsorption, that is, the metal ions on the mineral surface formed chemical bonds with the coordination groups in salicylhydroxamic acid, resulting in hydrophobic chelation on the mineral surface. Similarly, He et al. [45] found that octyl hydroxamic acid (salt) is an effective collector for smithsonite, as the N and O atoms in the functional groups interact with the Zn sites on the surface of smithsonite, improving the hydrophobicity of the mineral surface.

Based on the isomerism principle, Zhu et al. [46] studied the collection performance of salicylaldoxime and benzohydroxamic acid on smithsonite. The results of the tests reveal that both salicylaldoxime and benzohydroxamic acid exhibited promising collection performances for smithsonite. When benzohydroxamic acid was employed as the collector, a higher flotation recovery of smithsonite was obtained than that of salicylaldoxime. This suggests that benzohydroxamic acid has a stronger ability to collect smithsonite than salicylaldoxime. To explain the variation in the collection ability of the two reagents, a quantum chemistry method was applied to calculate the net charges of the bonding atoms of the benzohydroxamic acid and salicylaldoxime functional groups with metal ions. The calculation results indicate that the net charge of the bonding atoms between the functional groups in benzohydroxamic acid and the metal ions on the mineral surface was higher than that of the bonding atoms between the functional groups in salicylaldoxime and the metal ions on the mineral surface. Therefore, the reaction between benzohydroxamic acid and the smithsonite surface is stronger than that with salicylaldoxime.

Liu et al. [47] investigated the floatability and separation performance of smithsonite and calcite using benzohydroxamic acid as a collector and calcium lignosulfonate as a depressant, with the collector added first and the depressant added second. As illustrated in Figure 4a, benzohydroxamic acid can bind to the active sites on the surface of smithsonite and prevent the interaction between calcium lignosulfonate and Zn sites through spatial steric effects, resulting in a hydrophobic smithsonite surface. Furthermore, benzohydroxamic acid weakly interacts with the calcite surface, leaving enough space for calcium lignosulfonate to attach, resulting in a hydrophilic calcite surface. Wang et al. [61] used molecular dynamics simulation to simulate the chemisorption configuration of benzohydroxamic acid on the surface of smithsonite, as shown in Figure 4b. The distance between adjacent Zn atoms on the cleavage surface (101) of smithsonite crystal is 4.653 Å. Benzhydroxamic acid can be adsorbed onto the surface of smithsonite through chelation between the two O atoms in its head group and the Zn atoms, forming a five-membered ring. The weak interaction between benzohydroxamic acid and the calcite surface may be attributed to the low matching ability between the O–O distance of benzohydroxamic acid and the O–O distance of the carbonate group in calcite [62]. In addition, because Zn^{2+} belongs to the transition metal ion and Ca^{2+} belongs to the main group metal ion, the stability constant of the Zn-hydroxamate salt formed by the reaction of hydroxamate with Zn^{2+} is 7.31 [63], and the stability constant of the Ca-hydroxamate salt formed by the reaction of hydroxamate with Ca^{2+} is 2.4 [25]. Therefore, hydroxamate easily interacts with smithsonite containing Zn^{2+} on the surface, and the chelate that is formed is more stable [64,65].

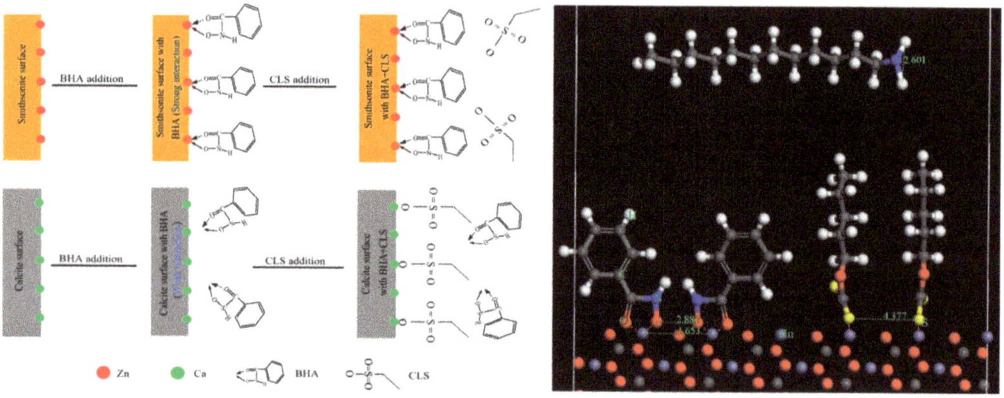

(a) (b)

Figure 4. (**a**) Schematic of the interaction of benzohydroxamic acid with mineral surface at pH 8.5 [47]; (**b**) adsorption configuration and spatial measurement [61].

Based on the hydroxylamine method, Zeng et al. [48] synthesized phenylpropene hydroxamic acid using methyl cinnamate, hydroxylamine, and anhydrous ethanol at 50 °C. Flotation separation tests were conducted on smithsonite and calcite, and it was found that, at a pH of 7 and a dosage of 50 mg/L of phenylpropene hydroxamic acid, the recovery of smithsonite reached 90.5%. Furthermore, the artificial mixed ore flotation test resulted in a Zn concentrate with a Zn grade of 35.1% and a Zn recovery of 81.4%. The X-ray photoelectron spectroscopy (XPS) analysis results indicate that phenylpropene hydroxamic acid interacts with the smithsonite surface to form a phenylpropene hydroxamic acid–smithsonite complex. The quantum chemistry calculations further reveal that phenylpropene hydroxamic acid has a higher highest occupied molecular orbital (HOMO) value than benzohydroxamic acid, making it easier for phenylpropene hydroxamic acid to provide electrons when reacting with smithsonite. Additionally, the higher ClogP value of phenylpropene hydroxamic acid compared with benzyl hydroxamic acid suggests that phenylpropene hydroxamic acid possesses better hydrophobic properties, which can effectively improve the surface hydrophobicity of the minerals.

The CF reagent, a metal-chelating collector developed by the Beijing General Research Institute of Mining and Metallurgy, is mainly composed of N-nitroso-N-phenylhydroxylammonium salt and has been applied in the flotation of wolframite and scheelite [66]. Tan et al. [67–69] utilized CF as a collector along with depressants such as sodium hexametaphosphate and a zinc sulfate salted water glass to effectively separate smithsonite from calcite, dolomite, limonite, and quartz through sulfidization–flotation under normal temperature and natural pH conditions. The O− and O= in CF could serve as chelating reagents with metal ions on the surface of smithsonite to form O–O five-membered ring complexes. Furthermore, Zhu et al. [49] found that cupferron exhibited good flotation recovery of smithsonite in both single-mineral and artificial mixed ore flotation tests, resulting in the effective separation of smithsonite from calcite and quartz. The theoretical analysis suggested that cupferron could adhere to the surface of smithsonite the most by forming insoluble Zn salts via the O atoms in the oxime group and the Zn ions on the surface of smithsonite, whereas a small amount of cupferron molecules may be physically adsorbed onto the surface of smithsonite.

Phosphate groups possess exceptional selective chelating abilities [70–72]. Phosphate esters are anionic surfactants used in industrial production and generally consist of monophosphates, diphosphates, and a small amount of phosphoric acid. Sun et al. [50] discovered that dodecyl phosphate ester potassium demonstrated ideal collection capabilities for smithsonite and could achieve an optimal flotation index for smithsonite under weakly alkaline conditions, with a flotation recovery of 90% under optimal conditions. The infrared spectrum (IR) and XPS analyses demonstrated that dodecyl phosphate ester potassium reacted with Zn atoms on the surface of smithsonite, producing Zn–O bonds by releasing H^+ in –OH, thus existing on the mineral surface through chemical adsorption (Figure 5). Moreover, compared to the common dodecylamine collector, dodecyl phosphate exhibits enhanced selectivity. Dodecylamine adheres to the surfaces of smithsonite and calcite simultaneously, whereas dodecyl phosphate ester potassium preferentially attaches to the surface of smithsonite, providing the potential for the effective separation of smithsonite and gangue minerals [73].

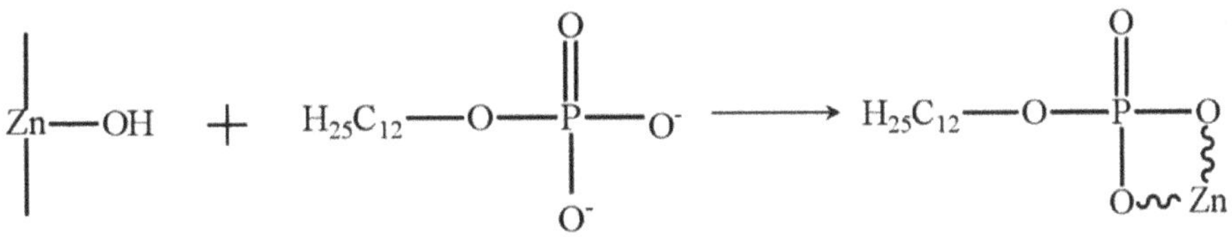

Figure 5. Model of dodecyl phosphate ester potassium adsorbed on the surface of smithsonite [50].

Qin et al. [51] conducted a flotation separation test of hemimorphite ore and quartz using di-(2-ethylhexyl) phosphoric acid as a collector and found that effective separation could be achieved within the pH range of 6–10. The anionic group in di-(2-ethylhexyl) phosphoric acid can interact with the Zn sites on the surface of hemimorphite and form a chemical bond, whereas the adsorption capacity on the surface of quartz is weak, leading to selective adsorption on the surface of the hemimorphite ore and the selective hydrophobic flotation of hemimorphite. Figure 6 shows that di-(2-ethylhexyl) phosphoric acid was chemically adsorbed on the surface of hemimorphite by forming a Zn–O bond between the O atom of the hydroxyl group and the Zn atom on the surface of the hemimorphite, resulting in the generation of a hydrophobic Zn(II)-di-(2-ethylhexyl) phosphoric acid surface complex on the mineral surface. Furthermore, di-(2-ethylhexyl) phosphoric acid reacts weakly with the Si atoms on the quartz surface, enabling the flotation separation of hemimorphite and quartz.

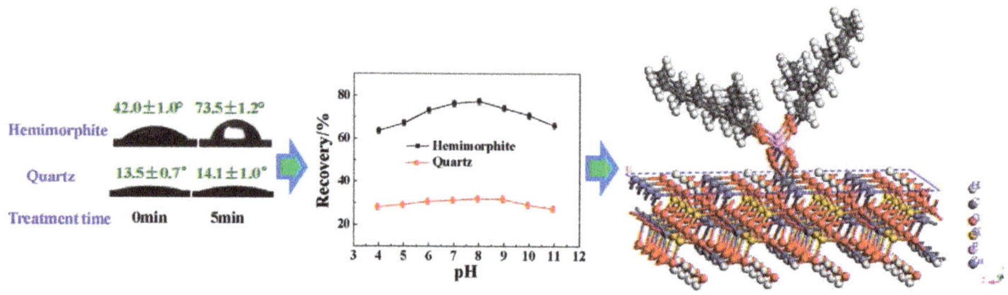

Figure 6. Selective flotation separation of hemimorphite from quartz under 1.0×10^{-4} mol L^{-1} di-(2-ethylhexyl) phosphoric acid and its hydrophobic mechanism on hemimorphite {1 1 0} surface [51].

N-lauroylsarcosine is a biosurfactant with amide and carboxyl groups. Because N-lauroylsarcosine can react with heavy metal ions via a complexation reaction, it was used to remove heavy metal ions from wastewater [52]. Based on the relationship between the group bond strength and complexation stability constant, Zhao et al. [53] designed and synthesized N-Lauroylsarcosine and N-dodecylglycine as flotation collectors of zinc oxide minerals using the bond valence model and density functional theory. The flotation results show that both have good collection performances for hemimorphite; however, the collection performance of N-lauroylsarcosine was better than that of N-dodecyl glycine. The mechanisms of action of N-lauroylsarcosine and N-dodecylglycine on the surface of hemimorphite are shown in Figure 7a. The polar groups in N-lauroylsarcosine and N-dodecylglycine can coordinate with the Zn (II) ions on the surface of hemimorphite through carboxyl and carbonyl groups, forming a seven-membered ring chelation in minerals. N-lauroylsarcosine has a better collection performance than N-dodecylglycine because it has two stable Zn (II)–C=O bonds in an aqueous solution, while the Zn (II)–C=O bond in N-dodecylglycine can break in an aqueous solution. Furthermore, Jia et al. [54] studied the adsorption mechanism of N-lauroylsarcosine on hemimorphite and quartz surfaces and found that N-lauroylsarcosine can interact with the Zn and H sites on the surface of hemimorphite to form a stable double-tooth chelation and be adsorbed on the surface of hemimorphite (Figure 7b), which enhances the surface hydrophobicity of hemimorphite and achieves the selective separation of hemimorphite and quartz.

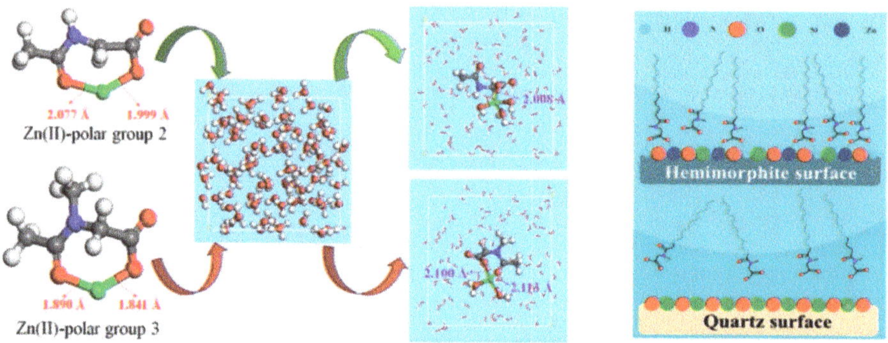

(a) (b)

Figure 7. (**a**) Configuration of Zn (II) complexes with N-lauroylsarcosine (group 2) and N-dodecylglycine (group 3) in aqueous solution (O—red; N—blue; C—gray; H—white; and Zn—green) [53]; (**b**) adsorption model of N-lauroylsarcosine on hemimorphite surface [54].

3.2. N–O Chelating Collectors

8–Hydroxyquinoline is a common metal chelating reagent that can react with many metal ions to form water-insoluble precipitates; therefore, it can interact with the surface of metal minerals and change their surface properties [55]. Rinelli et al. [56] used 8-hydroxyquinoline as a collector and found that the floatability of smithsonite was significantly improved under natural pH conditions. As shown in Figure 8, the N and O atoms in 8−hydroxyquinoline can form a five-membered ring chelate with the Zn atoms on the surface of smithsonite. To evaluate the characteristic adsorption of 8–hydroxyquinoline on the surface of smithsonite, Gabriela et al. [57] studied the effects of temperature and reagent concentration on the adsorption rate of 8–hydroxyquinoline on the surface of smithsonite using kinetics and equilibrium adsorption experiments. The results show that the temperature had no significant effect on the reaction rate. In addition, the adsorption activation energy of 8−hydroxyquinoline on the surface of smithsonite showed that the reaction occurred in the diffusion layer at the liquid phase interface.

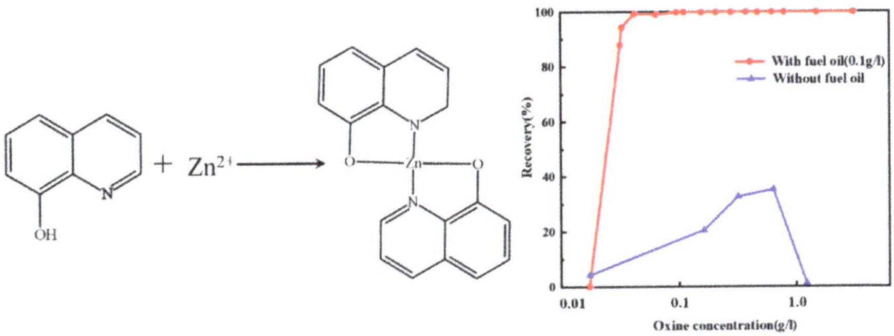

(a) (b)

Figure 8. (**a**) Reaction of 8−hydroxyquinoline with Zn^{2+} of smithsonite; (**b**) flotation recovery of smithsonite in Hallimond tube as a function of 8−hydroxyquinoline concentration at pH 7 [56].

Qiu et al. [74] studied a modified alkylamine chelating collector for smithsonite and achieved ideal separation indicators for the industrial production of zinc oxide ores. Alkylamine-chelating collectors have the advantages of good selectivity, strong collection performance, and a low dosage. Their structures contain groups that can chelate with the mineral surface and do not contain other groups that can be hydrated; therefore, they exhibit strong collecting performance. Based on the structural characteristics of zinc oxide minerals, Chen et al. [75] developed a new collector, C08, which could selectively collect zinc oxide minerals. The collector contains chelating functional groups composed of O, N, and P atoms that can strongly chelate Zn ions on the mineral surfaces to promote the hydrophobic flotation of zinc oxide minerals.

3.3. S–S and S–N Chelating Collectors

Mercaptobenzothiazole is a heterocyclic compound containing a thiazole ring and a sulfhydryl group that can chelate metal sites on the mineral surface and thus plays a role in collection [25]. Marabini et al. [58] studied the change rule of the floatability of smithsonite and cerussite when 6n-propoxy-mercapto-benzothiazole was used as the collector. The flotation test results show that under alkaline conditions, the collection capacity of 6n-propoxy-mercapto-benzothiazole on cerussite was stronger than that on smithsonite; however, under acidic conditions, the collection capacity of 6n-propoxy-mercapto-benzothiazole on smithsonite was stronger than that on cerussite. Therefore, when 6n-propoxy-mercapto-benzothiazole is used as the collector of lead–zinc oxide minerals, the flotation separation of minerals can be realized by controlling the pH of the pulp to adjust the floatability difference of the lead–zinc oxide minerals. In addition, the selectivity of aminothiophenol organic collectors for Zn in the flotation of lead–zinc minerals was evaluated using a statistical method. The results show that aminothiophenol had good collecting ability for Zn minerals, with the formation of ionic bonds between Zn and S, and had weak coordination bonds between N and Zn (Figure 9) [59].

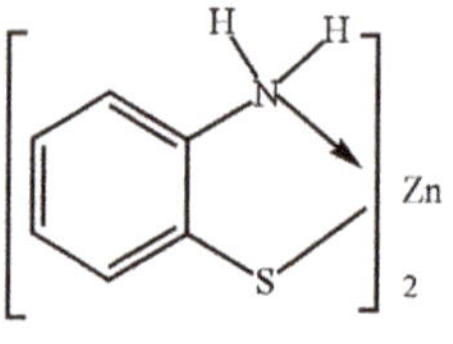

Figure 9. Reaction of aminothiophenol with Zn^{2+} of smithsonite [59].

4. Chelating Regulators

4.1. Chelating Activators

Chelating reagents can increase the number of active sites for the physical and chemical adsorption of collectors on mineral surfaces, thereby promoting collector adsorption and enhancing the hydrophobicity of the mineral surfaces [76]. Chelating reagents as activators used in the flotation of zinc oxide minerals can be classified into two categories based on their mechanisms of action. The first category involves the formation of hydrophobic chelates with metal ions on the mineral surface, which are then adsorbed onto the mineral surface, creating preliminary hydrophobic hydration of the mineral and promoting collector adsorption. The O–O and N–O chelating reagents in this category include salicylaldoxime, alpha-benzoic oxime, anthranilic acid, 8−hydroxyquinoline, and α-nitroso-β-naphthol. The second category involves the chelation of metal ions onto the mineral surface to form water-soluble chelates. These chelates enhance the dissolution of metal ions on the mineral surface and promote the multilayer adsorption of collectors onto the mineral surface, thereby activating the mineral. The chelating reagents in this category include ethylenediamine, ethylenediaminetetraacetic acid (EDTA), and xylenol orange. Figure 10 provides an overview of various chelating activators used in the flotation of zinc oxide ores.

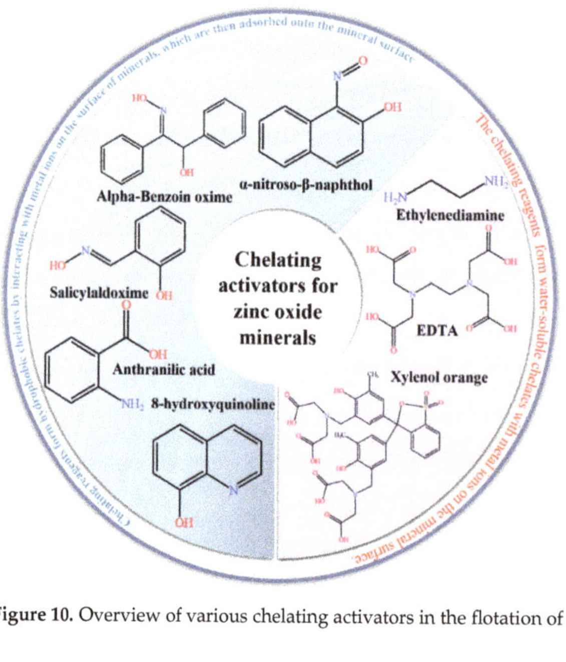

Figure 10. Overview of various chelating activators in the flotation of zinc oxide ores.

Wang et al. [77] studied the activation effects of salicyl aldoxime on smithsonite flotation. The results show that smithsonite can be activated by salicylaldoxime without sulfidization, and the flotation of smithsonite can be realized by the direct use of xanthate. The activation effect of salicylaldoxime may occur through the =NOH and -OH groups in its molecular structure and through Zn^{2+} on the surface of smithsonite, forming insoluble hydrophobic chelates on the mineral surface. This causes the mineral surface to be initially hydrophobic, destroying the hydration film on the mineral surface, and then promoting the adsorption of xanthate on the mineral surface. Wu et al. [78] studied the activation effects of salicylaldoxime, alpha-benzoic oxime, anthranilic acid, and 8–hydroxyquinoline on the flotation of smithsonite when dodecylamine was used as the collector. The results indicate that the flotation recovery of smithsonite increased with the increasing concentrations of salicylaldoxime, alpha-benzoic oxime, anthranilic acid, and 8-hydroxyquinoline. The four organic chelating reagents activated the flotation of smithsonite (Figure 11). Conversely, the N atom in the chelating activator and the O atom on the hydroxyl group can coordinate with the Zn ion on the surface of smithsonite to form an electrically neutral chelate with a cyclic structure. The chelate is attached to the surface of the mineral, which not only produces preliminary hydrophobic hydration, but also reduces the positive charge on the surface of the mineral, which is beneficial for the adsorption of amine collectors on the surface of smithsonite. Additionally, the chelate formed by the chelating activator on the surface of smithsonite can combine with amine molecules or amine ions through H bonds to promote the adsorption of amine collectors on the surface of smithsonite to achieve the purpose of activating smithsonite flotation.

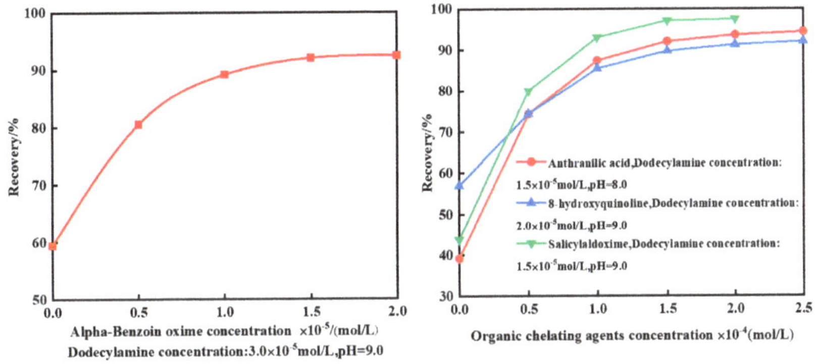

Figure 11. Effect of concentrations of alpha-benzoic oxime, anthranilic acid, 8—hydroxyquinoline, and salicylaldoxime on floatability of smithsonite [78].

α-Nitroso-β-naphthol is a chelating reagent with multiple active groups, which can form water-insoluble chelates with Cu^{2+}, Fe^{2+}, Co^{2+}, Ni^{2+}, Zn^{2+}, and other metal ions [79]. It was reported that α-nitroso-β-naphthol can be used as an activator in the flotation system of hemimorphite when octadecylamine and benzohydroxamic acid are used as collectors [80]. The activation of α-nitroso-β-naphthol can not only reduce the dosage of flotation reagents, but also promote the flotation separation of target minerals and gangue minerals [78]. It was proven via IR, ultraviolet (UV) spectroscopy, and group electronegativity calculations that α-nitroso-β-naphthol bonds with Zn^{2+} on the mineral surface through N and O atoms to form a five-membered ring chelate (Figure 12), which not only enhances the hydrophobicity of the mineral surface, but also promotes the adsorption of the collector on the mineral surface. In addition, the probability of α-nitroso-β-naphthol forming a five-membered ring product with the surface of hemimorphite through N and O atoms in the molecule exceeds the probability of forming other chelates; therefore, it is beneficial to promote the flotation separation of minerals [80,81].

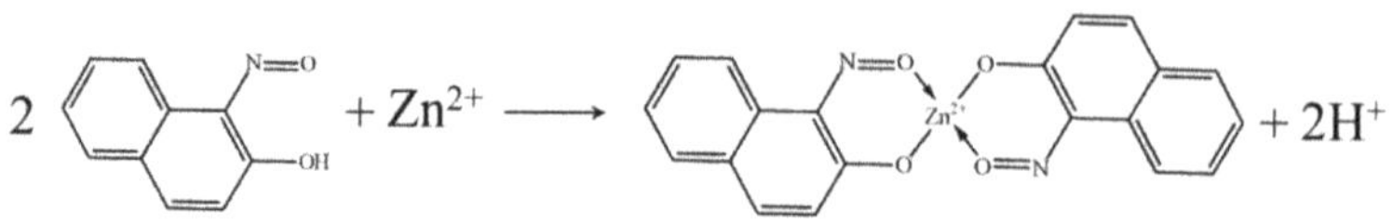

Figure 12. Reaction between α-nitroso-β-naphthol and Zn^{2+} on the hemimorphite surface [79].

After crushing and grinding, the Zn ions exposed on the surface of the smithsonite usually exist in two chemical states: charge-unsaturated Zn^{2+} and $CO_3{}^{2-}$ saturated Zn^{2+}. Wang et al. [82] and Wu et al. [78] proposed that the activation mechanism of ethylenediamine on smithsonite occurred through two paths. Owing to the high chemical activity of Zn^{2+} with an unsaturated charge, it easily reacts with ethylenediamine to form a water-soluble chelate, so that Zn^{2+} dissolves from the surface of the mineral, enhancing the negative charge on the surface of smithsonite and thereby facilitating the adsorption of cationic collectors on the surface of the mineral. Conversely, saturated Zn^{2+} has weak chemical activity, and the chelate formed after the reaction with ethylenediamine does not easily break away from the mineral surface and enter the liquid phase. The newly formed chelate can interact with the amine collector through H bonding to promote multilayer adsorption of the amine collector on the mineral surface, thereby increasing the adsorption amount of the collector on the mineral surface and realizing smithsonite activation flotation. Zhang et al. [83] studied the effect of ethylenediaminetetraacetic acid disodium calcium on the floatability of smithsonite in a dodecylamine system under different conditions. The results show that ethylenediaminetetraacetic acid disodium calcium could promote the dissolution of Zn^{2+} on the surface of smithsonite, reduce the potential on the surface of minerals, enhance the collecting ability of dodecylamine on smithsonite, and improve the flotation recovery of smithsonite. Similarly, Yang et al. [84] found that xylenol orange can form water-soluble chelating ions with Zn ions on the surface of hemimorphite, which promote the multilayer adsorption of amine ions on the surface of hemimorphite to activate its flotation.

4.2. *Chelating Depressants*

Because of the excellent selectivity of chelating reagents and the high stability of metal chelates formed via complexation with metal ions, chelating reagents with specific structures can be used as depressants for mineral flotation [85,86]. Chelating depressants mainly achieve the selective depression of minerals through three forms of action in the flotation process [87]: (1) chelating depressants interact with activation metal ions in the pulp to form stable chelates, thereby hindering the adsorption of metal ions on the mineral surface and eliminating the activation of activated ions on mineral flotation; (2) chelating depressants interact with the metal ions on the mineral surface to generate hydrophilic products attached to the mineral surface and hinder the adsorption of the collector, thereby enhancing the hydrophilicity of the mineral surface; and (3) chelation depressants react with the activation metal ions on the mineral surface to the pulp, thereby reducing the reactivity of the mineral surface.

Calcite is a typical gangue mineral in the flotation of zinc oxide ores. Because of its similar crystal structure and surface properties of smithsonite, it is difficult to effectively separate smithsonite from calcite during the flotation process, which significantly affects the quality of the Zn concentrate. Phytic acid is an organic phosphate compound that is extracted from plant seeds. Its molecular structure is shown in Figure 13. The carbon ring of a phytic acid molecule contains 12 free hydroxyl groups and six phosphate groups [88]. Therefore, phytic acid exhibits strong chelating ability for multivalent metal ions such as Ca^{2+} [89,90]. Chen et al. [91] found that phytic acid, as a depressant, can facilitate the flotation separation of smithsonite and calcite. At a pH value of 9.5, phytic acid can reduce the floatability of calcite from more than 94% to 10% and has little effect on the recovery of smithsonite. The results of the surface analysis show that phytic acid was easily chelated with Ca^{2+} exposed and adsorbed on the calcite surface, which hindered the adsorption of the collector on the surface of calcite, whereas only a slight amount of phytic acid was adsorbed on the smithsonite surface, thus achieving the selective depression of calcite (Figure 13). In addition, based on the difference in the crystal structure, the Zn–O bond length in the smithsonite crystal is 2.11 Å [92], and the Ca–O bond length in the calcite crystal is 2.36 Å [36]. In general, the shorter the bond length, the higher the bond energy. Compared with the Zn−O bond in smithsonite, the Ca–O bond length of calcite is longer, and its stability is lower; therefore, the surface of calcite has higher reactivity, which means it is more likely to chelate with the phosphate group [88]. Similarly, from the perspective of surface solubility, the solubility products of smithsonite and calcite were 1.46 $\times 10^{-10}$ and 4.96×10^{-9}, respectively. The solubility product of calcite is much higher than that of smithsonite, indicating that there are more free metal ions on the calcite surface, providing sufficient active sites for the action of phosphate groups on the calcite surface. In addition, in the smithsonite–calcite flotation system, the pH of the pulp was controlled at approximately 9.5. At this pH range, the calcium component in the pulp solution mainly exists as Ca^{2+} and $CaOH^{+}$ [93], whereas the Zn component mainly exists in the form of $Zn(OH)_2$ (aq) [94]. Compared to Zn $(OH)_2$ (aq), positively charged Ca^{2+} and $CaOH^{+}$ are more likely to interact with negatively charged phosphate groups [95].

Organophosphonic acids contain one or more $PO(OH)_2$ groups. They are also used in flotation owing to their stable C–P bonds and strong chelation with metal ions. The phosphonic acid groups (directly connected to carbon atoms), including –OH, $-CH_2$, and –COOH, and other groups in organic phosphonates can form three-dimensional bicyclic or polycyclic structures when chelated with Ca^{2+} [96]. Chen et al. [97] used amino trimethylene phosphonic acid as a depressant in the flotation system of smithsonite–calcite and found that a large amount of amino trimethylene phosphonic acid can chelate with the Ca^{2+} sites on the surface of calcite and be chemically adsorbed on the surface of calcite, thereby depressing the adsorption of sodium oleate on the surface of calcite (Figure 13). In the smithsonite flotation system, the amount of amino trimethylene phosphonic acid adsorbed on the surface of smithsonite was significantly lower than that on the surface of calcite, which almost did not affect the subsequent adsorption of sodium oleate on the surface of smithsonite. Chen et al. [97] believed that, in addition to the difference in the chelating properties of amino trimethylene phosphonic acid on the surface of the two minerals due to the different chemical compositions of the mineral surfaces, the distance between the O–O bond in the $PO(OH)_2$ group of amino trimethylene phosphonic acid and the distance of the Ca–Ca bond on the surface of calcite is better than that of the Zn–Zn bond on the surface of smithsonite, that is, the surface structure of calcite makes it easier for amino trimethylene phosphonic acid to chelate with the metal sites on the mineral surface and form a cyclic chelate [98].

Polyepoxysuccinic acid is a biodegradable and low-cost organic acid, and its molecular structure is shown in Figure 13. Liu et al. [99] found that the flotation separation of smithsonite and calcite could be achieved using sodium oleate as the collector and polyepoxysuccinic acid as a depressant. Polyepoxysuccinic acid can react with Ca^{2+} on the calcite surface to form a hydrophilic chelate and reduce the adsorption sites of sodium oleate on the calcite surface, thereby enhancing the hydrophobicity of the calcite surface.

Quartz is one of the typical gangues in zinc oxide ores. In the flotation of quartz with sodium oleate as a collector, because of the lack of active sites for sodium oleate on the quartz surface, it is not easy to directly collect quartz with sodium oleate, showing weak floatability. However, owing to the high solubility of smithsonite, Zn^{2+} in smithsonite dissolves from the mineral surface to the pulp during the grinding and flotation processes [100]. The interaction between Zn^{2+} in the pulp and the quartz surface enhances the reactivity of the quartz surface with the collector, which greatly improves the floatability of quartz and increases the difficulty of flotation separation of smithsonite and quartz. Based on the characteristic that the Zn active sites on the quartz surface can form soluble complexes with chelating reagents, Wang et al. and Zhao et al. [101–103] investigated the selective depression mechanism of 1-hydroxyethylidene-1,1-diphosphonic acid, sodium polyaspartate, and tetrasodium iminodisuccinate on a smithsonite–quartz flotation system when sodium oleate was used as the collector. The structures of 1-hydroxyethylidene-1,1-diphosphonic acid, sodium polyaspartate, and tetrasodium iminodisuccinate are shown in Figure 14. The results show that 1-hydroxyethylidene-1,1-diphosphonic acid, sodium polyaspartate, and tetrasodium iminodisuccinate could chelate the Zn sites on the surface of quartz and transfer them to the solution, resulting in a decrease in the adsorption of sodium oleate on the surface of quartz, which effectively weakened the floatability of quartz. Because of the different crystal structures of smithsonite and quartz, the chemical environments and reactivities of the Zn sites on the surfaces of the two minerals were different. The reactions of 1-hydroxyethylidene-1,1-diphosphonic acid, sodium polyaspartate, and tetrasodium iminodisuccinate with the Zn sites on the surface of smithsonite were weak. Therefore, they did not have significant effects on the floatability of smithsonite to realize the flotation separation of smithsonite and quartz. The conceivable separation mechanism using chelating depressants is depicted in Figure 14.

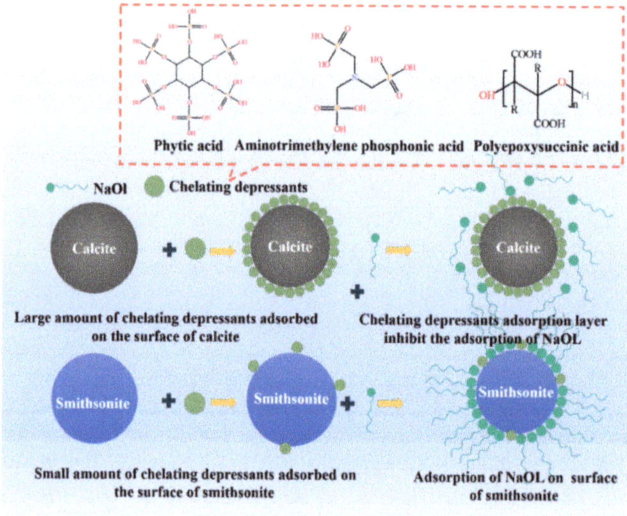

Figure 13. Structure of chelating depressants and its interaction model with smithsonite and calcite [90,97,99].

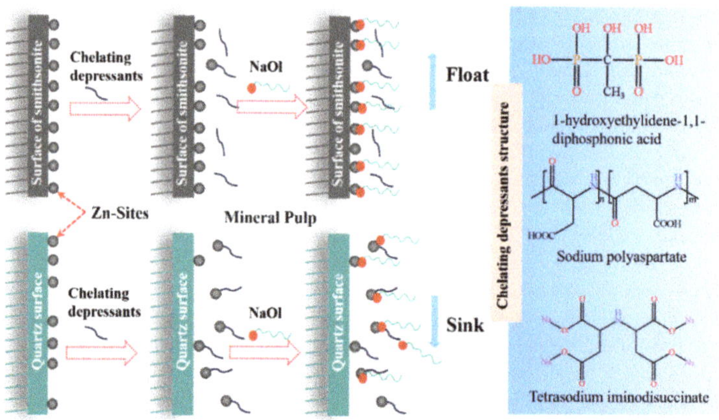

Figure 14. Structure of chelating depressants and its interaction model with smithsonite and quartz [101–103].

5. Conclusions and Outlook

Flotation is an effective and economical method for the treatment of zinc oxide ores. However, it is difficult to achieve the efficient flotation recovery of zinc oxide minerals in industrial production because of the difficulty of the flotation of zinc oxide minerals and the difficulty of mineral separation owing to the similar surface properties of the target minerals and gangue minerals. Recently, the development of novel high-selectivity zinc oxide flotation reagents has become a research hotspot. Various chelating reagents have been used in the flotation of zinc oxide minerals, effectively promoting the progress of zinc oxide ore flotation technology. In this study, combined with the crystal structure and surface properties of typical zinc oxide and gangue minerals, the research progress on chelating reagents and mechanisms in zinc oxide ore flotation is reviewed. The conclusions are as follows:

(1) When chelating reagents are used as collectors for zinc oxide minerals, the molecular structure usually includes polar chelating groups composed of O, N, P, and S and hydrophobic non-polar hydrocarbon groups. In the flotation process of zinc oxide minerals, chelating collectors can react with Zn^{2+} on the surface of zinc oxide minerals to form hydrophobic chelates, thereby improving the surface floatability. The pH range of 5–10 is optimum for the interaction of the chelating collector with the zinc oxide mineral surface. In this pH range, the main components on the surfaces of smithsonite and hemimorphite are Zn^{2+} and $ZnOH^{+}$, which can provide sufficient active sites for the adsorption of chelating collectors on the mineral surface. Pulp pH can affect the chemical state of the active site on the zinc oxide mineral surface and gangue surface, and then affect the reaction between the mineral surface and chelating reagents. Therefore, the selectivity of the chelating reagent can be enhanced via pH adjustment, and the flotation separation of zinc oxide minerals and gangue can be realized. In addition, the collection capacity of chelating reagent is also related to the chelating group activity, the hydrophobic chain length, and other factors.

(2) Chelating activators are commonly used in the flotation of zinc oxide ores when amine cations are used as collectors. In addition to the chelating group, the structure of the chelating activator usually contains a benzene ring or a shorter alkyl group. It primarily activates zinc oxide minerals by initially enhancing the hydrophobicity of the zinc oxide mineral surface and promoting the dissolution of the mineral surface.

(3) Chelating depressants used in the flotation of zinc oxide ores are mostly derived from scale depressants in the field of environmental protection. The molecular structure of chelating depressants usually contains polar functional groups ($-PO(OH)_2$, –COOH, etc.) that can chelate metal ions on the mineral surface, and the hydrophilic groups (–OH, etc.) in chelating depressants can enhance the mineral surface's hydrophilic property. In addition, chelating depressants can achieve the selective depression of gangue by adsorbing onto the surface of gangue to hinder the adsorption of collectors and dissolve the active ions on gangue surface to reduce their reactivity with collectors.

The use of chelating reagents in the flotation of zinc oxide ores is encouraging, and ideal flotation results have been obtained in laboratory flotation tests. However, the application of these new chelating reagents for the flotation of zinc oxide ores in industrial production faces many challenges, including flotation index, cost, and environmental pressure. Further research should be carried out on the combination of different chelating reagents, and the combination of new chelating reagents and traditional reagents. Examples include the combination of chelating collectors with traditional collectors such as xanthate, dodecylamine, and sodium oleate, and the combination of chelating depressants/activators and inorganic depressants/activators. In addition, it is necessary to further investigate the chelating effect of chelating groups on minerals with similar physical and chemical surface properties to better understand the interaction mechanism between chelating reagents and mineral surfaces. Understanding the chelation effect between the chelating group of the reagent and the metal ions on the surface of the mineral can provide theoretical guidance for the selection or design of new selective collectors or depressants/activators for zinc oxide ores.

Author Contributions: Z.S.: Data Curation, Formal Analysis, Investigation, Methodology, Software, Visualization, Writing—Original Draft, and Writing—Review and Editing. S.W.: Investigation, Methodology, Project Administration, Resources, Supervision, Validation, and Writing—Review and Editing. G.H.: Conceptualization, Data Curation, Formal Analysis, Funding Acquisition, Investigation, Methodology, Project Administration, Resources, Supervision, Validation, and Writing—Review and Editing. Q.F.: Formal Analysis, Funding Acquisition, Investigation, Methodology, Project Administration, Resources, Supervision, Validation, Visualization, and Writing—Review and Editing. All authors have read and agreed to the published version of the manuscript.

Funding: This work was supported by the National Natural Science Foundation of China (Grant No. 52264026), the Yunnan Science and Technology Leading Talent Project (Grant No. 202305AB350005), and Yunnan Fundamental Research Projects (Grant No. 202301AW070018).

Data Availability Statement: Not applicable.

Conflicts of Interest: The authors declare no conflict of interest.

References

1. Liu, C.; Feng, Q.; Zhang, G.; Ma, W.; Meng, Q.; Chen, Y. Effects of lead ions on the flotation of hemimorphite using sodium oleate. *Miner. Eng.* **2016**, *89*, 163–167. [CrossRef]
2. Deng, R.; Wang, Y.; Duan, W.; Xing, D.; Hu, Y. Induced crystallization of Pb^{2+} on smithsonite surface during sulfidation-xanthate flotation. *Colloids Surf. A-Physicochem. Eng. Asp.* **2022**, *650*, 129576. [CrossRef]
3. Ejtemaei, M.; Irannajad, M.; Gharabaghi, M. Influence of important factors on flotation of zinc oxide mineral using cationic, anionic and mixed (cationic/anionic) collectors. *Miner. Eng.* **2011**, *24*, 1402–1408. [CrossRef]
4. Luo, Y.; Ou, L.; Chen, J. Experimental and computational study of the differential sulfidization mechanism of smithsonite and calcite. *Miner. Eng.* **2023**, *198*, 108102. [CrossRef]
5. Abkhoshk, E.; Jorjani, E.; Al-Harahsheh, M.S.; Rashchi, F.; Naazeri, M. Review of the hydrometallurgical processing of non-sulfide zinc ores. *Hydrometallurgy* **2014**, *149*, 153–167. [CrossRef]
6. Zhao, W.; Yang, B.; Yi, Y.; Feng, Q.; Liu, D. Synergistic activation of smithsonite with copper-ammonium species for enhancing surface reactivity and xanthate adsorption. *Int. J. Min. Sci. Technol.* **2023**, *33*, 519–527. [CrossRef]
7. Mehdilo, A.; Irannajad, M.; Zarei, H. Smithsonite Flotation from Zinc Oxide Ore using Alkyl Amine Acetate Collectors. *Sep. Sci. Technol.* **2014**, *49*, 445–457. [CrossRef]
8. Chen, A.; Zhao, Z.; Jia, X.; Long, S.; Huo, G.; Chen, X. Alkaline leaching Zn and its concomitant metals from refractory hemimorphite zinc oxide ore. *Hydrometallurgy* **2009**, *97*, 228–232. [CrossRef]
9. Hosseini, S.H.; Forssberg, E. Smithsonite flotation using potassium amyl xanthate and hexylmercaptan. *Miner. Process. Extr. Metall.* **2006**, *115*, 107–112. [CrossRef]
10. Liao, R.; Wen, S.; Feng, Q.; Deng, J.; Lai, H. Activation mechanism of ammonium oxalate with pyrite in the lime system and its response to flotation separation pyrite from arsenopyrite. *Int. J. Miner. Metall. Mater.* **2023**, *30*, 271–282. [CrossRef]
11. Shen, Z.; Wen, S.; Wang, H.; Miao, Y.; Wang, X.; Meng, S.; Feng, Q. Effect of dissolved components of malachite and calcite on surface properties and flotation behavior. *Int. J. Miner. Metall. Mater.* **2023**, *30*, 1297–1309. [CrossRef]
12. Salum, M.J.G.; de Araujo, A.C.; Peres, A.E.C. The role of sodium sulphide in amine flotation of silicate zinc minerals. *Miner. Eng.* **1992**, *5*, 411–419. [CrossRef]
13. Chen, Y.; Liu, M.; Chen, J.; Li, Y.; Zhao, C.; Mu, X. A density functional based tight binding ($DFTB^+$) study on the sulfidization-amine flotation mechanism of smithsonite. *Appl. Surf. Sci.* **2018**, *458*, 454–463. [CrossRef]
14. Pereira, C.A.; Peres, A.E.C. Reagents in calamine zinc ores flotation. *Miner. Eng.* **2005**, *18*, 275–277. [CrossRef]
15. Feng, Q.; Zhao, W.; Wen, S. Ammonia modification for enhancing adsorption of sulfide species onto malachite surfaces and implications for flotation. *J. Alloys Compd.* **2018**, *744*, 301–309. [CrossRef]
16. Hosseini, S.H.; Forssberg, E. Physicochemical studies of smithsonite flotation using mixed anionic/cationic collector. *Miner. Eng.* **2007**, *20*, 621–624. [CrossRef]
17. Liu, Y.; Hu, X.; Wei, Z. Overview of Research on Flotation Reagents of Zinc Oxide Ore. *Conserv. Util. Miner. Resour.* **2011**, *1*, 51–55. (In Chinese)
18. Hosseini, S.H.; Forssberg, E. Adsorption studies of smithsonite flotation using dodecylamine and oleic acid. *Min. Met. Explor.* **2006**, *23*, 87–96. [CrossRef]
19. de Weldige, K.; Rohwerder, M.; Vago, E.; Viefhaus, H.; Stratmann, M. Adsorption of self-assembled monolayers of mercaptan on gold. *Fresenius' J. Anal. Chem.* **1995**, *353*, 329–332. [CrossRef]
20. Wu, Q.; Guo, X.; Shi, X.; Liu, Q.; Pei, J.; Sun, Y. Research progress on chelating reagents for flotation. *Ind. Miner. Process.* **2015**, *44*, 55–59. (In Chinese)
21. Liu, W.; Wei, D.; Zhou, D.; Zhu, Y.; Jia, C. Application of chelating collectors in flotation. *Met. Ore Dress. Abroad.* **2006**, *7*, 4–8+20. (In Chinese)
22. Marabini, A.M.; Ciriachi, M.; Plescia, P.; Barbaro, M. Chelating reagents for flotation. *Miner. Eng.* **2007**, *20*, 1014–1025. [CrossRef]

23. Wu, W.; Sun, C.; Zhu, Y. "1+1/4" Necessary and Sufficient Conditions of Organic Chelating Agents as Depressors or Activators. In Proceedings of the XXIV International Mineral Processing Congress, Beijing, China, 24–28 September 2008; pp. 1558–1562.
24. Liu, B. *Chelating Flotation Reagents*, 1st ed.; Metallurgical Industry Press: Beijing, China, 1982; pp. 9, 71–75, 170–182.
25. Jian, B. *Flotation Reagents*, 1st ed.; Metallurgical Industry Press: Beijing, China, 1985; pp. 289–292.
26. Fuerstenau, D.W.; Herrera-Urbina, R.; McGlashan, D.W. Studies on the applicability of chelating agents as universal collectors for copper minerals. *Int. J. Miner. Process.* **2000**, *58*, 15–33. [CrossRef]
27. Han, H.; Hu, Y.; Sun, W.; Li, X.; Cao, C.; Liu, R.; Yue, T.; Meng, X.; Guo, Y.; Wang, J.; et al. Fatty acid flotation versus BHA flotation of tungsten minerals and their performance in flotation practice. *Int. J. Miner. Process.* **2017**, *159*, 22–29. [CrossRef]
28. Liu, C.; Zhang, W.; Song, S.; Li, H. Study on the activation mechanism of lead ions in wolframite flotation using benzyl hydroxamic acid as the collector. *Miner. Eng.* **2019**, *141*, 105859. [CrossRef]
29. Angadi, S.I.; Sreenivas, T.; Jeon, H.-S.; Baek, S.-H.; Mishra, B.K. A review of cassiterite beneficiation fundamentals and plant practices. *Miner. Eng.* **2015**, *70*, 178–200. [CrossRef]
30. Wang, J.; Sun, Z.; Bai, J. Research on Crystal Anisotropy and Surface Properties of Smithsonite. *Conserv. Util. Miner. Resour.* **2021**, *41*, 1–6. (In Chinese)
31. Luo, Y.; Ou, L.; Zhang, G.; Chen, J.; Li, Y.; Shi, Q.; Zhu, B.; Xia, Y.; Chen, S.; Zhang, Z.; et al. The effect of surface vacancy on adsorption of HS on smithsonite (101) surface: A DFT study. *Colloid Surf. A-Physicochem. Eng. Asp.* **2021**, *624*, 126713. [CrossRef]
32. Ejtemaei, M.; Gharabaghi, M.; Irannajad, M. A review of zinc oxide mineral beneficiation using flotation method. *Adv. Colloid Interface Sci.* **2014**, *206*, 68–78. [CrossRef]
33. Irannajad, M.; Ejtemaei, M.; Gharabaghi, M. The effect of reagents on selective flotation of smithsonite–calcite–quartz. *Miner. Eng.* **2009**, *22*, 766–771. [CrossRef]
34. Ejtemaei, M.; Irannajad, M.; Gharabaghi, M. Role of dissolved mineral species in selective flotation of smithsonite from quartz using oleate as collector. *Int. J. Miner. Process.* **2012**, *114–117*, 40–47. [CrossRef]
35. Wang, Q.; Zhang, X.; Liu, D.; Cao, S.; Song, K.; Jing, M.; Li, K.; Wu, L.; Liu, R. Basic Characteristics of Hemimorphite and Its Transformation Mechanism with Na_2CO_3. *Minerals* **2018**, *8*, 143. [CrossRef]
36. Gao, Z.; Li, C.; Sun, W.; Hu, Y. Anisotropic surface properties of calcite: A consideration of surface broken bonds. *Colloid Surf. A-Physicochem. Eng. Asp.* **2017**, *520*, 53–61. [CrossRef]
37. Wang, J.; Zhang, Q.; Qiu, Y.; Li, L.; Ye, J.; Cui, W. The first principles of the crystal structure and active sites of calcite. *Chin. J. Eng.* **2017**, *39*, 487–493. (In Chinese)
38. Shi, T.; Zhang, Q. Effects of Crystal Structure and Surface Properties on Quartz Floatability. *Acta Mineral. Sin.* **2017**, *37*, 333–341. (In Chinese)
39. Crundwell, F.K. On the mechanism of the flotation of oxides and silicates. *Miner. Eng.* **2016**, *95*, 185–196. [CrossRef]
40. Kowalczuk, P.B. Flotation and hydrophobicity of quartz in the presence of hexylamine. *Int. J. Miner. Process.* **2015**, *140*, 66–71. [CrossRef]
41. Guo, W.; Zhu, Y.; Han, Y.; Li, Y.; Yuan, S. Flotation performance and adsorption mechanism of a new collector 2-(carbamoylamino) lauric acid on quartz surface. *Miner. Eng.* **2020**, *153*, 106343. [CrossRef]
42. Pattanaik, A.; Venugopal, R. Investigation of Adsorption Mechanism of Reagents (Surfactants) System and its Applicability in Iron Ore Flotation—An Overview. *Colloid Interface Sci. Commun.* **2018**, *25*, 41–65. [CrossRef]
43. Liu, B.; Wang, X.; Du, H.; Liu, J.; Zheng, S.; Zhang, Y.; Miller, J.D. The surface features of lead activation in amyl xanthate flotation of quartz. *Int. J. Miner. Process.* **2016**, *151*, 33–39. [CrossRef]
44. Zhao, J.; Zhu, J. Using salicylhydroxamic acid as collector in flotation of smithsonite and lead sulfate. *Nonferrous Met.* **1991**, *4*, 27–32. (In Chinese)
45. He, X.; Cheng, D.; Gong, Z. Study on flotation behavior of zinc oxide minerls with chelating collector-octyl hydroxamate. *J. Guangdong Non-Ferr. Met.* **1991**, *1*, 1–6. (In Chinese)
46. Zhu, J.; Wu, X. Synthesizing collector of oxide flotation with principle of isomerism. *Nonferrous Met.* **1990**, *3*, 32–3743. (In Chinese)
47. Liu, C.; Zhu, Y.; Huang, K.; Yang, S.; Liang, Z. Studies of benzyl hydroxamic acid/calcium lignosulphonate addition order in the flotation separation of smithsonite from calcite. *Int. J. Min. Sci. Technol.* **2021**, *31*, 1153–1158. [CrossRef]
48. Zeng, Y.; Yao, X.; Liu, G.; He, G.; Yu, X.; He, G.; Huang, Z.; Zhang, R.; Cheng, C. Flotation behavior and mechanism of phenylpropenyl hydroxamic acid for the separation of smithsonite and calcite. *J. Mol. Liq.* **2021**, *339*, 116893. [CrossRef]
49. Zhu, J.; Zhao, J. The experiments of cupferron as collector in the flotation of lead sulfate and smithonite. *J. Cent. South Univ. (Sci. Technol.)* **1991**, *22*, 522–528. (In Chinese)
50. Sun, Q.; Feng, Q.; Shi, Q. Adsorption mechanism of smithsonite by dodecyl phosphate ester potassium. *J. Cent. South Univ. Sci. Technol.* **2018**, *49*, 1845–1850. (In Chinese)
51. Qin, J.; Liu, G.; Fan, H.; Tan, W. The hydrophobic mechanism of di(2-ethylhexyl) phosphoric acid to hemimorphite flotation. *Colloid Surf. A-Physicochem. Eng. Asp.* **2018**, *545*, 68–77. [CrossRef]
52. Jia, K.; Yi, Y.; Ma, W.; Cao, Y.; Li, G.; Liu, S.; Wang, T.; An, N. Ion flotation of heavy metal ions by using biodegradable biosurfactant as collector: Application and removal mechanism. *Miner. Eng.* **2022**, *176*, 107338. [CrossRef]
53. Zhao, L.; Liu, W.; Duan, H.; Wang, X.; Fang, P.; Liu, W.; Zhou, X.; Shen, Y. Design and selection of flotation collectors for zinc oxide minerals based on bond valence model. *Miner. Eng.* **2021**, *160*, 106681. [CrossRef]

54. Jia, K.; Lu, Y.; Liu, J.; Cheng, S.; Liu, S.; Cao, Y.; Li, G. Selective flotation separation of hemimorphite from quartz using the biosurfactant sodium N-lauroylsarcosinate as a novel collector. *Miner. Eng.* **2023**, *198*, 108073. [CrossRef]
55. Chen, Z.; Yao, J.; Knudsen, T.Š.; Ma, B.; Liu, B.; Li, H.; Zhu, X.; Zhao, C.; Pang, W.; Cao, Y. Degradation of novel mineral flotation reagent 8-hydroxyquinoline by superparamagnetic immobilized laccase: Effect, mechanism and toxicity evaluation. *Chem. Eng. J.* **2022**, *432*, 134239. [CrossRef]
56. Rinelli, G.; Marabini, A.M. *Flotation of Zinc and Lead Oxide–Sulphide Ores with Chelating Agents;* X International Mineral Processing Congress: London, UK, 1973; pp. 493–521.
57. Oprea, G.M.; Michnea, A.M.; Mihali, C. Adsorption kinetics of 8-hydroxyquinoline on smithsonite. *Rev. Roum. Chim.* **2011**, *56*, 1021–1027.
58. Tan, X.; Li, C. Research progress on flotation of lead-zinc oxide ore at home and abroad (I). *Met. Ore Dress. Abroad.* **2000**, *37*, 7–14. (In Chinese)
59. Barbaro, M.P. Comparision of Pb-Zn selective collectors using statistical methods. *Miner. Eng.* **1999**, *12*, 356–366. [CrossRef]
60. Cai, J.; Deng, J.; Wang, L.; Hu, M.; Xu, H.; Hou, X.; Wu, B.; Li, S. Reagent types and action mechanisms in ilmenite flotation: A review. *Int. J. Miner. Metall. Mater.* **2022**, *29*, 1656–1669. [CrossRef]
61. Wang, Z.; Xu, L.; Wang, J.; Wang, L.; Xiao, J. A comparison study of adsorption of benzohydroxamic acid and amyl xanthate on smithsonite with dodecylamine as co-collector. *Appl. Surf. Sci.* **2017**, *426*, 1141–1147. [CrossRef]
62. Jiang, W.; Gao, Z.; Khoso, S.A.; Gao, J.; Sun, W.; Pu, W.; Hu, Y. Selective adsorption of benzhydroxamic acid on fluorite rendering selective separation of fluorite/calcite. *Appl. Surf. Sci.* **2018**, *435*, 752–758. [CrossRef]
63. Wu, H.; Yuan, J.; Zhou, K. Determination of Stability Constants of Peutylohydroxamic Acid (PHA) Heptylohydroxamic Acid (HHA) with Copper(II) Zinc(II) Nickel(II) and Manganese(II). *J. Nanchang Univ. Nat. Sci.* **1988**, *1988*, 96–102. (In Chinese)
64. Fuerstenau, M.C.; Harper, R.W.; Miller, J.D. Hydroxamate vs. Fatty Acid Flotation of Iron Oxide. *Soc. Min. Metall. Explor.* **1970**, *247*, 69–72.
65. Miller, J.D.; Wang, X.; Li, M. Bench Scale Flotation of Sedimentary Phosphate Rock with Hydroxamic Acid Collectors. In Proceedings of the Engineering Foundation Conference: Beneficiation of Phosphate III, St. Pete Beach, FL, USA, December 2001.
66. Xiao, Q.; Li, C.; Kang, G. Study on the flotation process of tungsten trunk by CF method in Shizhuyuan polymetallic ore. *Min. Metall.* **1996**, *3*, 26–32. (In Chinese)
67. Tan, X.; Li, C. Lead and zinc oxide ores flotation with CF as collector. *Nonferrous Met.* **2002**, *04*, 86–94. (In Chinese)
68. Tan, X.; Li, C. Study on the mechanism of interaction between CF collector and the surfaces of oxidized lead and zinc minerals I. study of adsorption, ζ-potential and IR spectra. *Min. Metall.* **2004**, *3*, 23–29. (In Chinese)
69. Tan, X.; Li, C. Study on mechanism of interaction between CF collector and surfaces of oxidized lead and zinc minerals II. study of group electronegativity of flotation agent and XPS tests. *Min. Metall.* **2005**, *1*, 24–28. (In Chinese)
70. Jiang, X.; Zhang, W.; Fan, R.; Zhang, Z.; Chen, S.; Pooley, S.; Yang, L.; Gao, Z. Improved flotation of chalcopyrite from galena and pyrite by employing Cu-affinity phosphate collector. *Miner. Eng.* **2023**, *197*, 108064. [CrossRef]
71. Marion, C.; Li, R.; Waters, K.E. A review of reagents applied to rare-earth mineral flotation. *Adv. Colloid Interface Sci.* **2020**, *279*, 102142. [CrossRef]
72. Wang, J.; Zhou, Z.; Gao, Y.; Sun, W.; Hu, Y.; Gao, Z. Reverse Flotation Separation of Fluorite from Calcite: A Novel Reagent Scheme. *Minerals* **2018**, *8*, 313. [CrossRef]
73. Liu, W.; Wang, Z.; Wang, X.; Miller, J.D. Smithsonite flotation with lauryl phosphate. *Miner. Eng.* **2020**, *147*, 106155. [CrossRef]
74. Qiu, Y.; Zhou, Y.; Tang, X.; Yang, M. Experimental research on flotation of zinc oxide minerals by using novel chelating collector E-5. *Nonferrous Met. Miner. Process.* **2007**, *4*, 43–4637. (In Chinese)
75. Cheng, J.; Kang, R.; Ke, B. Research on New Reagents Flotation of Refractory Oxided Zinc Ore. *Conserv. Util. Miner. Resour.* **2012**, *5*, 16–19. (In Chinese)
76. Feng, Q.; Yang, W.; Wen, S.; Wang, H.; Zhao, W.; Han, G. Flotation of copper oxide minerals: A review. *Int. J. Min. Sci. Technol.* **2022**, *32*, 1351–1364. [CrossRef]
77. Wang, L.; Len, E.; Liu, B. A Shudy on The Flotation of oxygen Zinc Minerals Using Salicylaldoxime as Activator and Xanthate as Collector. *J. Kunming Univ. Sci. Technol.* **1991**, *4*, 31–34. (In Chinese)
78. Wu, W.; Sun, C.; Zhu, Y. Study on the activating mechanism of five types of organic chelating agents for smithsonite. *Min. Metall.* **2007**, *16*, 16–21. (In Chinese)
79. Hua, Q.; Guo, H.; Wang, D.; Huang, Y.; Cao, Y.; Peng, W.; Fan, G. A new strategy for selective recovery of low concentration cobalt ions from wastewater: Based on selective chelating precipitation-flotation process. *J. Taiwan Inst. Chem. Eng.* **2022**, *141*, 104605. [CrossRef]
80. Zhou, W. Study on Activation Behavior of α-Nitroso-β-Naphthol in Flotation of Hemimorphite. Master's Thesis, Hunan University, Changsha, China, 2010.
81. Meng, W.; Mo, H.; Zhou, W.; Xu, F.; Xu, W.; Ou, E. Flotation of zinc oxide ore by using α-nitroso-β-naphthol as activator. *J. Hunan Univ. Nat. Sci. Ed.* **2017**, *44*, 107–111. (In Chinese)
82. Wang, Z. Effect and mechanism of organic activator in smithsonite flotation. *Nonferrous Met.* **1996**, *3*, 13–18. (In Chinese)
83. Zhang, W.; Bi, H.; Zhang, J.; Li, G.; Hao, H.; Han, C.; Wei, D.; Shen, Y. Effect of EDTA Na_2-Ca on the Pelagic Activity of Smithsonite in Dodecylamine System. *Conserv. Util. Miner. Resour.* **2018**, *3*, 124–129. (In Chinese)

84. Yang, Y.; Liu, B.; Len, E. Study on flotation of hemimorphite activated by xylenol orange. *Yunnan Metall.* **1992**, *21*, 35–38. (In Chinese)
85. El-bahi, A.; Taha, Y.; Ait-Khouia, Y.; Hakkou, R.; Benzaazoua, M. Advancing phosphate ore minerals separation with sustainable flotation reagents: An investigation into highly selective biobased depressants. *Adv. Colloid Interface Sci.* **2023**, *317*, 102921. [CrossRef]
86. Wang, X.; Xie, R.; Liu, J.; Zhu, Y. Selectively separating scheelite from fluorite by using Nitrilotri (methylphosphonic acid) as an efficiency depressant to modify the surface properties of fluorite. *Colloid Surf. A-Physicochem. Eng. Asp.* **2023**, *674*, 131878. [CrossRef]
87. Wu, W.; Sun, C.; Zhu, Y. Application of Organic Chelating Depressants in Flotation. *Nonferrous Met.* **2006**, *4*, 81–85. (In Chinese)
88. Chen, W.; Feng, Q.; Zhang, G.; Yang, Q. Investigations on flotation separation of scheelite from calcite by using a novel depressant: Sodium phytate. *Miner. Eng.* **2018**, *126*, 116–122. [CrossRef]
89. Tsao, G.T.; Zheng, Y.; Lu, J.; Gong, C. Adsorption of heavy metal ions by immobilized phytic acid. *Appl. Biochem. Biotechnol.* **1997**, *63–65*, 731–741. [CrossRef] [PubMed]
90. Cui, X.; Li, Q.; Li, Y.; Wang, F.; Jin, G.; Ding, M. Microstructure and corrosion resistance of phytic acid conversion coatings for magnesium alloy. *Appl. Surf. Sci.* **2008**, *255*, 2098–2103. [CrossRef]
91. Chen, Y.; Guo, X.; Chen, Y. Using phytic acid as a depressant for the selective flotation separation of smithsonite from calcite. *Sep. Purif. Technol.* **2022**, *302*, 122104. [CrossRef]
92. Liu, J.; Zeng, Y.; Ejtemaei, M.; Nguyen, A.V.; Wang, Y.; Wen, S. DFT simulation of S-species interaction with smithsonite (0 0 1) surface: Effect of water molecule adsorption position. *Results Phys.* **2019**, *15*, 102575. [CrossRef]
93. Wang, L.; Gao, H.; Song, S.; Zhou, W.; Xue, N.; Nie, Y.; Feng, B. The depressing role of sodium alginate in the flotation of $Ca2^{+}$-activated quartz using fatty acid collector. *J. Mol. Liq.* **2021**, *343*, 117618. [CrossRef]
94. Bai, S.; Li, C.; Fu, X.; Ding, Z.; Wen, S. Promoting sulfidation of smithsonite by zinc sulfide species increase with addition of ammonium chloride and its effect on flotation performance. *Miner. Eng.* **2018**, *125*, 190–199. [CrossRef]
95. Wang, L.; Lyu, W.; Zhou, W.; Zhang, H. The role of sodium phytate in the flotation separation of smithsonite from calcite. *Miner. Eng.* **2022**, *187*, 107775. [CrossRef]
96. Gao, Z.; Wang, C.; Sun, W.; Gao, Y.; Kowalczuk, P.B. Froth flotation of fluorite: A review. *Adv. Colloid Interface Sci.* **2021**, *290*, 102382. [CrossRef]
97. Chen, Y.; Tang, X. Selective flotation separation of smithsonite from calcite by application of amino trimethylene phosphonic acid as depressant. *Appl. Surf. Sci.* **2020**, *512*, 145663. [CrossRef]
98. Liu, C.; Zhang, W.; Song, S.; Li, H.; Liu, Y. Flotation separation of smithsonite from calcite using 2-phosphonobutane-1,2,4 tricarboxylic acid as a depressant. *Powder Technol.* **2019**, *352*, 11–15. [CrossRef]
99. Liu, C.; Wang, X.; Yang, S.; Ren, Z.; Li, C.; Hu, Z. Utilization of polyepoxysuccinic acid as a green depressant for the flotation separation of smithsonite from calcite. *Miner. Eng.* **2021**, *168*, 106933. [CrossRef]
100. Duarte, G.M.P.; Lima, R.M.F. Quartz and Hematite Activation by Zn, Ca and Mg Ions in the Cationic Flotation Route for Oxidized Zinc Ore. *Miner. Process Extr. Metall. Rev.* **2021**, *43*, 720–727. [CrossRef]
101. Wang, M.; Zhang, G.; Zhao, L.; Chen, Y.; Liu, D. Utilization of 1-hydroxyethylidene-1,1-diphosphonic acid to selectively separate smithsonite from zinc ions activated quartz. *Miner. Eng.* **2022**, *182*, 107585. [CrossRef]
102. Zhao, L.; Zhang, G.; Wang, M.; Zheng, S.; Li, B. Selective separation of smithsonite from quartz by using sodium polyaspartate as a depressant. *Colloid Surf. A-Physicochem. Eng. Asp.* **2022**, *644*, 128840. [CrossRef]
103. Wang, M.; Zhang, G.; Zhao, L.; Chen, Y.; Liu, D.; Li, C. Application of eco-friendly tetrasodium iminodisuccinate for separation of smithsonite from zinc ions activated quartz. *Miner. Eng.* **2022**, *181*, 107545. [CrossRef]

minerals

MDPI

Review

Sources, Performance and Mechanisms of Metal Ions in the Flotation Process of Copper, Lead, and Zinc Ores: A Review

Bihan Wei [1], Yuqiang Mao [1], Liang Wang [1,2], Peilun Shen [1], Han Wang [1,*] and Dianwen Liu [1,2,*]

[1] State Key Laboratory of Complex Nonferrous Metal Resources Clean Utilization, Key Laboratory of Green Separation and Enrichment of Strategic Mineral Resources, Faculty of Land Resource Engineering, Kunming University of Science and Technology, Kunming 650093, China; 18313560620@163.com (B.W.); cumtmyq@163.com (Y.M.); wangliangkust@126.com (L.W.); peilunshen@kust.edu.cn (P.S.)

[2] China Southwest United Graduate School, Kunming 650092, China

* Correspondence: 20240077@kust.edu.cn (H.W.); dianwenliu1@kust.edu.cn (D.L.); Tel.: +86-13759104687 (D.L.)

Abstract: In mineral processing operations, besides target minerals, slurries often contain various metal ions, including common ones with different valence states such as Pb^{2+}, Cu^{2+}, Fe^{2+}, and Fe^{3+}. These metal ions originate from multiple sources during the flotation process, including the dissolution of co-occurring metal minerals during crushing and grinding, the addition of flotation reagents, and the flotation water. Investigators have long recognized that metal ions significantly affect mineral flotation behavior. Due to physical and chemical interactions, some ions in the slurry will interact with target mineral. If these ions form hydrophobic substances on the mineral surface or increase the active sites between the mineral and collectors or sulfide agent, they will have a positive impact on the flotation process. Conversely, if they produce hydrophilic substances or deplete added collectors and sulfides, they negatively impact mineral enrichment. Meanwhile, metal ions can regulate the electrostatic repulsion between reagents and mineral surfaces in the slurry system, which has a certain impact on the flotation results. This study provides a comprehensive overview of the sources of metal ions in flotation, explores their adsorption characteristics on mineral surfaces, and examines their impact on the flotation process. It provides a theoretical basis for improving mineral flotation processes in the presence of metal ions.

Keywords: flotation; metal ions; copper; lead; zinc ores

1. Introduction

Copper, lead, and zinc metals are primarily extracted from copper, lead, and zinc ores, respectively [1–3]. Copper is renowned for its outstanding electrical and thermal conductivity, which makes it extensively utilized across various industries, including electrical systems, construction materials, transportation, telecommunications, and electrical engineering [4]. Lead is predominantly utilized in vehicle production, building materials, batteries, and the electronics industry [5]. Additionally, lead serves as a key component in the steel industry [6]. Zinc finds widespread use in steel metallurgy, construction, electronics, and chemical engineering, and is employed in manufacturing anti-corrosion coatings, batteries, alloys, and more [7]. Its chemical properties also make zinc an important auxiliary agent in metallurgy, enhancing the performance of alloys [8].

Flotation is the main processing method for sulide ores Cu, Pb, Zn ores, which achieves effective separation and enrichment of target minerals [9–11]. Given the diminishing availability of high-grade mineral resources and increasing environmental regulations, improving resource utilization is a pressing challenge. During mineral processing, metal ions commonly appear in the slurry of co-occurring metal minerals during crushing and grinding, wear of grinding media, addition of flotation reagents, and water quality [12–14]. These metal ions can either activate or inhibit mineral flotation [15,16]. Improving the positive effect of metal ions on copper, lead, and zinc minerals, or alleviating their adverse effects, has the potential to greatly enhance flotation recovery efficiency.

As shown in Figures 1 and 2, researchers have extensively studied the influence mechanisms of different metal ions on different types of flotation systems of copper lead zinc ores, and have published many studies. Metal ions may have vastly different effects on the flotation of sulfide copper ore under different conditions, and metal ions can have an impact on both the sulfide flotation of oxidized copper ore and the flotation of sulfide oxidized copper ore; Cu^{2+} is the most common activating ion for sphalerite, while Pb^{2+}'s effect on sphalerite varies depending on the conditions. Both ions can have a positive effect on zinc oxide in sulfide flotation systems; Some metal ions can inhibit the flotation of sulfide lead ore under strong alkaline conditions, while excessive ions not only consume sulfurizing agents but also change the pH of the ore slurry, leading to unstable flotation during sulfide flotation of oxidized lead ore. This study conducts a comprehensive review of prior research to establish a theoretical foundation for optimizing the separation processes of copper, lead, and zinc.

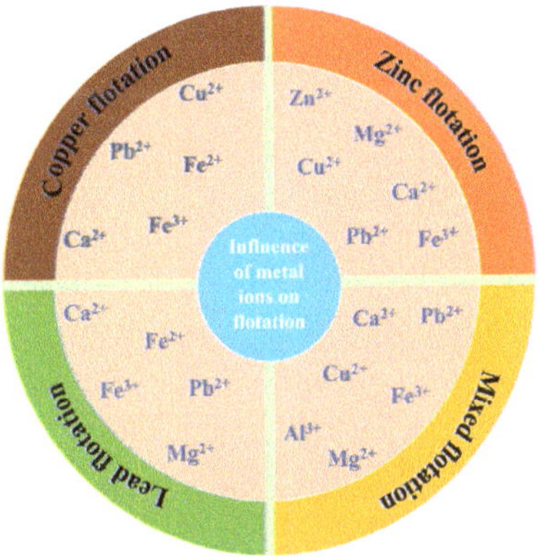

Figure 1. The types of metal ions that affect the flotation of copper, lead, zinc, and mixed minerals, respectively.

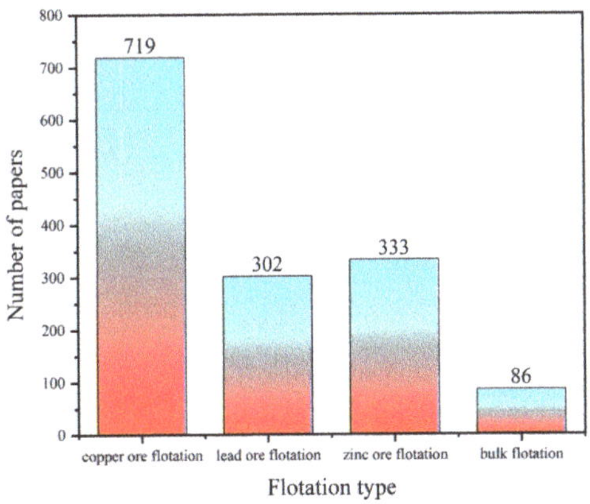

Figure 2. Number of papers on the impact of metal ions on copper, lead, and zinc flotation in the past decade.

2. Sources of Metal Ions

2.1. Mineral Factors

2.1.1. Lattice Impurities

Owing to mineralization and geological processes, impurity atoms or ions can become incorporated into the crystal lattice, causing distortions or changes in the crystal structure. Most of these impurities are metal ions, which can impact flotation when they enter the slurry [17,18]. For example, the presence of iron (Fe) impurities within the crystal lattice of sphalerite can significantly influence its flotation efficiency by modifying both the chemical characteristics and semiconductor properties of the sphalerite surface [19].

2.1.2. Mineral Dissolution

The dissolution of both useful and gangue minerals releases metal ions into the flotation system. The dissolution process is predominantly affected by the surface properties of the minerals, their solubility, and the composition of the slurry solution. Oxidized and silicate minerals are particularly susceptible to surface dissolution due to their distinct cations in their crystal structures [20–22]. Some minerals contain complex compositions and soluble metal salts embedded as tiny particles in the ore. When these ores enter the slurry system, they dissolve and release metal ions. This can influence the flotation process for both target and non-target minerals, which may result in higher consumption of reagents and a decline in the quality of the flotation environment [23–26].

2.1.3. Release of Fluid Inclusions

Fluid inclusions are formed when diagenetic and mineralizing fluids are encapsulated during the process of mineral formation. These inclusions are frequently observed within natural mineral crystals [27–29]. As shown in Table 1, during mineral processing, ores are subjected to crushing and grinding, which releases these fluid inclusions into the slurry. The released inclusions may contain metal ions that can impact flotation. For example, the release of Pb^{2+}, Cu^{2+}, and Fe^{3+} from fluid inclusions in natural galena and chalcopyrite during grinding represents a significant new source of metal ions [30,31].

Table 1. Concentration of ions released from fluid inclusions during 8 min of mineral grinding (10^{-6} mol/L).

	Ca^{2+}	Pb^{2+}	Cu^{2+}	Fe^{3+}	Mg^{2+}
Chalcopyrite [30]	-	-	0.21	0.80	-
Sphalerite [32]	1.53	-	-	-	0.22
Galena [31]	-	4.32	-	-	-

2.2. Pulp System

2.2.1. Flotation Reagents

The various reagents added during the flotation process are one of the main sources of metal ions in the slurry. As shown in Table 2, when adding reagents to adjust the slurry system or mineral surface, a large amount of metal ions are often passively or actively introduced. When using lime to adjust the pH value of the slurry, a large amount of Ca^{2+} will also enter the slurry, which may inhibit the flotation of some minerals under alkaline conditions $ZnSO_4$ and $CuSO_4$ are commonly used as inhibitors and activators for sphalerite, the introduction of Zn^{2+} and Cu^{2+} can influence the floatability of minerals in flotation systems [33,34].

Table 2. Common metal salts and their industrial applications.

Drug Name	Metal Ions Entering the Slurry	Application
Cupric sulfate ($CuSO_4{\cdot}5H_2O$)	Cu^{2+}	Activators for sphalerite, pyrite, etc.
Zinc sulfate ($ZnSO_4{\cdot}7H_2O$)	Zn^{2+}	Sphalerite inhibitor
Ferrous sulfate ($FeSO_4{\cdot}7H_2O$)	Fe^{2+}	Inhibiting sphalerite with cyanide [35]
Ferric chloride ($FeCl_3$)	Fe^{3+}	Barite inhibitor
Aluminum trichloride ($AlCl_3$)	Al^{3+}	Inhibiting calcite [36]
Lead nitrate ($Pb(NO_3)_2$)	Pb^{2+}	Activated sphalerite with copper sulfate [37,38]

2.2.2. Flotation Water Quality

The flotation process consumes a large amount of water. To adhere to the principles of sustainable development, environmental protection, and improving water resource utilization, concentrators usually use dehydration and water circulation systems to reuse flotation water [39,40]. Due to mineral dissolution, equipment corrosion, and the addition of chemicals, there may be a large amount of metal ions in the recovered water from mineral processing. If not handled properly, they may re-enter the process and have adverse effects on flotation [41,42].

Various mineral processing plants employ different qualities of water for flotation [43]. In regions with limited freshwater, seawater flotation is widely adopted [44–47]. In certain remote areas, high-salinity water, which is rich in Ca^{2+}, Mg^{2+} for flotation [48,49]. Rotting animals and plants naturally degrade and subsequently aggregate to produce a large amount of humic acid, which exists in natural water. Using water containing these substances for flotation will result in fluctuations in mineral flotation indicators [50]. The quality of water utilized in the flotation process can have a profound effect on its overall efficiency and effectiveness [51].

2.2.3. Grinding Media

During the grinding process, equipment such as ball mills or grinders can introduce metal ions from the materials used. The collision between minerals and grinding media releases Fe^{3+} and Cr^{3+}, which can influence oxidation-reduction reactions on the surfaces of both minerals and grinding media, thereby impacting the slurry environment. Water, commonly used for transportation and cooling in grinding, may contain soluble iron compounds, leading to the introduction of Fe^{3+} into the slurry system [52,53].

3. Influence of Metal Ions on Target Minerals

Copper, lead, and zinc deposits usually contain various minerals, with only specific target minerals being valuable for development and utilization. In copper, lead, and zinc ores, target minerals are those that contain copper, lead, or zinc and possess a defined chemical composition and fixed crystal structure [54–56].

3.1. *Influence of Metal Ions on Copper Mineral Flotation*

Copper sulfide is the primary type of copper ore, including chalcopyrite, chalcocite, and bornite. Chalcopyrite, the main source of copper, often coexists with pyrite, sphalerite, galena, and various non-metallic gangue minerals to form sulfide polymetallic ores [57–59]. Chalcopyrite exhibits good natural floatability in neutral and weakly alkaline media [60,61]. However, the introduction of certain metal ions into the slurry can alter the mineral's surface properties, affecting its hydrophilicity and complicating its separation from other minerals [62–64].

3.1.1. Effect of Metal Ions on the Flotation of Copper Sulfide Ore

Activated Effect of Flotation

Metal ions interact with flotation reagent to form reaction sites that interact with mineral surfaces, thereby promoting the flotation of copper sulfide. Cu^{2+} can activate chalcopyrite through two mechanisms. First, Cu^{2+} reacts with xanthate ions in the slurry to form copper xanthate, which adsorbs onto the sulfur atoms on the chalcopyrite surface, creating a hydrophobic surface. Cu^{2+} interacts with the active sites S on the surface of chalcopyrite to form Cu-S species and form a hydrophobic surface [65]. Carboxymethyl cellulose (CMC) is an effective depressant of silicate and carbonate minerals, used to suppress minerals such as quartz, talc, and calcite. Under specific conditions, it can also serve as a depressant of chalcopyrite. Under alkaline conditions, the electrostatic repulsion between CMC molecules and the surface of chalcopyrite limits its inhibitory effect, while the introduction of positively charged Fe^{3+} ions can effectively alleviate this electrostatic repulsion, thereby enhancing the adsorption of CMC. At a pH of 4, Fe^{3+} ions interact with CMC molecules, resulting in chemical adsorption onto the chalcopyrite surface, moreover, the content of positively charged iron species on the surface of chalcopyrite is higher than that under alkaline conditions, which is more conducive to the adsorption of CMC [66].

Depressed Effect of Flotation

Some metal ions can form hydrophilic substances on mineral surfaces, negatively impacting flotation by consuming regulators and collectors. Fe^{3+} inhibits the flotation of chalcopyrite under strong alkaline conditions because the iron sites on the mineral surface preferentially oxidize and hydrophilic substances, hindering the adsorption of xanthate and reducing the floatability of chalcopyrite [67]. When using Na_2CS_3 as a chalcopyrite inhibitor, the addition of Cu^{2+} enhances the inhibitory effect. $CS_3{}^{2-}$ hydrolyzes to produce OH^- and reacts with Cu^{2+} to produce $Cu(OH)_2$, which covers the chalcopyrite surface and impedes xanthate adsorption [68].

In summary, metal ions exhibit varying impacts on the flotation of copper sulfide contingent upon the specific conditions present. During flotation, interactions between metal ions, mineral surfaces, and reagents may involve a series of ion exchange and electrochemical reactions. Further exploration and verification of the reaction products and mechanisms of different metal ions are needed.

3.1.2. Influence of Metal Ions on Copper Oxide Flotation

As resources continue to be depleted, there has been a significant decline in copper sulfide resources, and the mining and recover of copper oxide have become a new research direction [69,70]. Typical oxidized copper ores encompass malachite, azurite, chrysocolla, and cuprite [71–74]. The methods for recovering copper oxide typically encompass several techniques: direct flotation, which employs chelating reagents or fatty acids as collectors; sulfidation flotation, in which xanthate is utilized as a collector following sulfidation; and activation flotation, using chelating agents, ammonium/amine salts, metal ions, and oxidants for activation to recycle copper oxide ore [75].

Activated Effect of Flotation

Ionic bonds are the main bonds formed by the surface fracture of copper oxide, with extremely strong hydrophilicity [76]. Direct flotation of oxidized copper ore using fatty acid collectors is generally inefficient. However, the addition of metal ions can compress the double layer on the mineral surface, facilitating the aggregation of mineral particles. For instance, in the pH range of 6–11, Ca^{2+} forms $Ca(OH)^+$ species that reduce the double-layer repulsion between malachite particles. This reduction increases particle aggregation and improves the recovery rate [77].

The ability of sulfur ions to penetrate the water shell and adsorb onto mineral surfaces determines the degree of sulfurization processes on oxidized mineral surfaces. The sulfidation process of oxidized copper ore is influenced by numerous factors, leading to instability and potentially poor flotation efficiency [78–80]. Certain metal ions can improve the availability of active sites on surfaces of mineral, facilitating the adsorption of subsequent sulfide ions or collectors, thereby acting as activators and promoting the sulfide flotation of oxidized copper ore [33,81–83]. For example, pre-treatment of malachite with Pb^{2+} ions significantly improves the adsorption of sulfide ions and xanthates onto its surface [84]. Low concentrations of Cu^{2+} promote the sulfidation process of malachite by enhancing the reactivity of the Cu-O species present on the mineral surface and promoting the generation of Cu-S species [82]. Pb^{2+} adsorbs onto the surface of azurite, forming a Pb-O structure that strongly attracts sulfide ions and generates Pb-S component on the mineral surface. This results in a fully sulfided azurite surface, which enhances the adsorption of xanthates [85,86]. Yang et al. proposed a flotation method for stepwise activating silicon chrysocolla with Pb^{2+}. The introduction of Pb^{2+} ions both prior to and following the sulfidation process results in an increased formation of Pb-S compounds and additional Pb^{2+} reaction sites on the surface of chrysocolla. This enhancement greatly improves the reactivity of the sulfidized product [87].

For flotation of oxidized ore after sulfidation, adding specific metal ions can produce active substances on the surface of mineral that enhance the adsorption of xanthates. Treatment of sulfurized malachite with Pb^{2+} leads to the formation of S^{2-}, $S_n{}^{2-}$, and $SO_n{}^{2-}$, which generate stable Pb-S component on the surface of malachite, thereby promoting the sulfurization of the surface of malachite [88]. Pb^{2+} and $Pb(OH)^+$ adsorb onto the surface of sulfide azurite and react with sulfide ions, resulting in more Cu(II) species being reduced to Cu(I) species and forming highly active, low solubility polysulfide species [89]. As shown in Figure 3, an appropriate concentration of Cu^{2+} forms a Cu(I)-S component with S ions on the sulfide product of malachite, increasing the activity of the sulfide product and improving flotation [90].

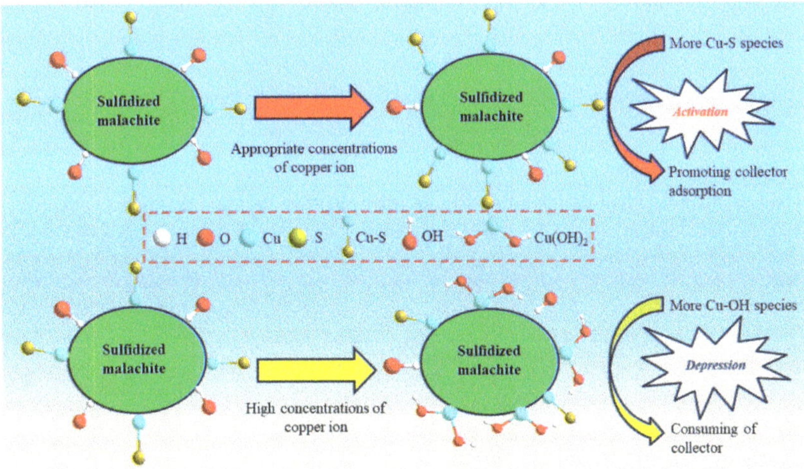

Figure 3. A mechanistic model of the effect of copper ions on the flotation of sulfurized malachite [90].

Appropriate concentration of ammonium salt can improve the sulfurization environment during copper oxide sulfide flotation [24]. Moreover, the synergistic activation of copper oxide by metal ions and ammonium salts can further enhance the formation of sulfide substances on the mineral surface and promote sulfide flotation. In the flotation process of malachite, $Cu(NH_3)_n{}^{2+}$ component is formed under the synergistic activation system of copper and ammonium, providing additional Cu sites on the surface of malachite and promoting the adsorption of sulfur ions and collectors [91]. The introduction of $(NH_4)_2SO_4$ prior to the sulfidation process serves to suppress the formation of copper sulfide colloids while alleviating the adverse effects associated with an excess of sodium sulfide. This is attributed to the increased solubility of azurite due to the formation of $Cu(NH_3)_n{}^{2+}$, sulfur ions that should have formed copper sulfide colloids and dissolved copper ions form hydrophobic films on the surface of azurite [92,93].

Cu^{2+} or hydroxy copper component on the surface of oxidized copper ore can form cyclic compounds and adsorb on the surface of minerals through covalent or coordination bonds with functional groups in chelating collectors [94,95]. Chelating agents may also serve as activators by chelating Cu^{2+} ions on the surface of copper oxide, resulting in the formation of hydrophobic chelate complexes. This interaction significantly improves the adsorption of collectors [96]. The chelation between triethanolamine and Cu^{2+} on the malachite surface, along with the weak dissolution effect on the mineral surface, increases malachite's surface activity, and strengthens the chemical adsorption of xanthate and sodium sulfide on malachite and facilitates the formation of a relatively tight multilayer adsorption of xanthate [97]. As shown in Figure 4, the use of DMTD(2,5-dimercapto-1,3,4-thiadiazol) as an activator for malachite pre-treatment significantly improves flotation efficiency. The interaction between sulfur in DMTD and Cu^{2+} on the malachite surface forms a Cu-S material, which benefits xanthate adsorption [98]. Li et al. studied the mechanism of action of hydroxyoctylphosphonic acid as a collector on the surface of malachite, their findings indicate that HPA chelates with Cu^{2+} on the surface of malachite to form a Cu-O-P-O cyclic structure or Cu-P linear adsorption. This reaction forms a Cu-HPA complex adsorption layer on the malachite surface [99].

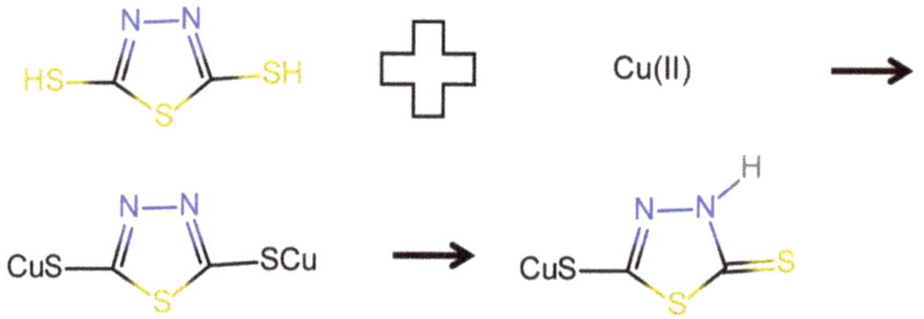

Figure 4. Mechanism of DMTD on the surface of malachite [98].

Depressed Effect of Flotation

Ca^{2+} is a prevalent metal ion present in the flotation systems employed for oxidized ores. Research conducted by Li et al. has demonstrated that at pH levels below 6, Ca^{2+} can interact with octyl hydroxamate, leading to a significant impact on flotation performance. Conversely, when the pH surpasses 11, hydrophilic species and Ca^{2+} precipitation form on the surface of malachite, which not only reduces hydrophobicity but also hinders the adsorption of collectors [77]. The presence of Fe^{2+} or Fe^{3+} ions impedes the adsorption of sulfide ions on the surface of copper oxide by forming Fe-OH species within the slurry. This reaction consequently diminishes the activity of Cu(I)-S active component on the mineral surface, lead to inadequate sulfidation of the mineral. Furthermore, Fe-OH species can also form on the surface of sulfide products associated with malachite, which significantly compromises the activity of these products and cause a decrease in flotation recovery rate. Additionally, Fe^{3+} ions may interact with the Cu(I)-S species on the mineral surface, causing excessive oxidation of the sulfide products. This interaction weakens the mineral surface's adsorption capacity for collectors and further inhibits flotation efficiency [25,100]. Excessive Cu^{2+} in the slurry mainly exists as $Cu(OH)_2$ on the malachite surface, enhancing the mineral's hydrophilicity. However, on the surface of sulfurized malachite, Cu-(OH)/$Cu(OH)_2$ components are generated, which reduce the reaction activity of sulfurized products and are not conducive to reactions with xanthates [88,101].

As shown in Table 3, metal ions play a significant role in influencing the flotation process of both copper oxide and sulfide minerals. Specifically, these active metal ions can enhance the activity of active sites on the surface of copper oxide, facilitating interactions with sodium sulfide. Moreover, when metal ions are combined with ammonium salts, they can further improve the sulfidation degree of copper oxide ores. In the context of direct flotation, metal ions can engage with chelating agents, resulting in the formation of hydrophobic chelates that contribute to improved flotation recovery rates. Nonetheless, it is important to note that under certain conditions, metal ions may induce the formation of hydrophilic species on the surface of mineral. This alteration can impede the flotation process and may inadvertently lead to the depletion of collectors or adjusters used in the flotation process.

Table 3. Effect of typical metal ions on copper mineral flotation.

	Cu^{2+}	Pb^{2+}	Fe^{2+}	Fe^{3+}
Chalcopyrite	Activated effect of flotation	Activated effect of flotation	-	Depressed effect of flotation under strong alkaline conditions
Malachite	Low concentration activate flotation, high concentration depress flotation	Activated effect of flotation	Depressed effect of flotation	Depressed effect of flotation
Azurite	-	Activated effect of flotation	-	-
Chrysocolla	-	Activated effect of flotation	-	-
Cuprite	-	-	-	Depressed effect of flotation

3.2. *Influence of Metal Ions on Zinc Mineral Flotation*

Zinc ore is categorized into two primary types: zinc sulfide ore and zinc oxide ore. Notably, zinc sulfide ore possesses substantially larger reserves and exhibits a broader geographic distribution in comparison to zinc oxide ore.

3.2.1. Influence of Metal Ions on the Flotation of Zinc Sulfide Ore

In zinc sulfide, zinc primarily exists as sphalerite or iron sphalerite. Zinc sphalerite (ZnS) is the most significant zinc mineral, constituting over 90% of the raw materials for zinc smelting. However, pure zinc minerals are rare in nature, and sphalerite is typically found in association with minerals such as lead, copper, and precious metals [37].

Activated Effect of Flotation

Sphalerite demonstrates excellent floatability in acidic environments and can be activated by metal ions to enhance flotation in alkaline conditions. The mechanism of Cu^{2+} ions on sphalerite has emerged as a significant area of research. It is generally believed that under acidic conditions, Cu^{2+} undergoes a substitution reaction with ZnS in sphalerite to form Cu-S material; however, this Cu-S material is only metastable, with prolonged oxidation, various copper-sulfur compounds with different oxidation states form on the surface of sphalerite, these copper–sulfur compounds interact more effectively with collectors, making sphalerite more hydrophobic [102,103]. The activation mechanism of sphalerite by Cu^{2+} in alkaline conditions is commonly understood to involve the formation of a $Cu(OH)_2$ layer on the surface of the mineral. This layer subsequently engages in an ion exchange process with the zinc present on the sphalerite surface when exposed to acidic conditions. This interaction ultimately leads to the development of stable Cu-S species and the generation of hydrophilic $Zn(OH)_2$ [33,104,105]. The mechanism of copper ion interaction under these two conditions is shown in Figure 5 and the steps are shown in Table 4.

Table 4. Reaction steps for Cu^{2+} activation of sphalerite.

Pulp pH Conditions	Reaction Steps
Acidic conditions	$(ZnS)ZnS + Cu^{2+} \rightarrow (ZnS)CuS + Zn^{2+}$ $yZnS + xCu^{2+} \rightarrow (y - x)ZnSxCu{+}S^{-} + xZn^{2+}$
Alkaline conditions	$Cu^{2+} + 2OH^{-} \leftrightarrow Cu(OH)_{2\,(S)}$ $Cu(OH)_{2\,(S)} + Zn^{2+}{}_{(aq)} \leftrightarrow (Zn,Cu)S_{(s)} + Zn^{2+} + 2OH^{-}$ $Cu(OH)_{2\,(S)} \leftrightarrow Cu^{2+}{}_{(aq)} + 2OH^{-}$ $Zn^{2+}{}_{(aq)} + 2OH^{-} \leftrightarrow Zn(OH)_{2\,(S)}$

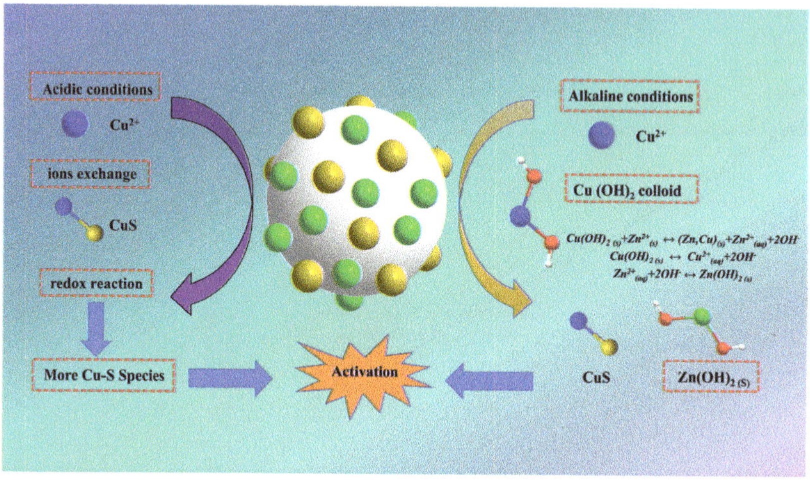

Figure 5. Exchange model of Cu^{2+} on the surface of sphalerite.

Under acidic and weakly alkaline conditions, Pb^{2+} can activate sphalerite in two ways. One view is that Pb^{2+} interacts with O to form -Zn-O-Pb bonds on the sphalerite surface, creating lead oxide material and facilitating adsorption [106]. Another view suggests that Pb^{2+} directly interacts with Zn^{2+} and S^{2-} on the sphalerite surface and adsorbs onto it. $ZnSO_4$ is commonly used as an inhibitor of sphalerite in industry, and Zn^{2+} forms $Zn(OH)_2$ colloids under alkaline conditions, which adhere to the sphalerite surface and impede the action of the surface and collector [107]. Under acidic conditions, Pb^{2+} can activate sphalerite that is inhibited by $ZnSO_4$, hydrolyzed lead species, especially Pb and Pb(OH), interact with the surface of sphalerite to form a-Zn-O-Pb complexes. This complex increases the reaction sites on the surface of sphalerite, thereby promoting the adsorption of the collector [38]. Hydrolyzed Cu^{2+} produces $Cu(OH)_2$, $Cu(OH)^{3-}$, and $Cu(OH)_4{}^{2-}$ species that adsorb onto the sphalerite surface inhibited by $ZnSO_4$. This creates new sites for xanthate adsorption, thereby relieving the inhibitory effect on sphalerite and activating its flotation [34]. Fe^{2+} can activate the flotation of sphalerite within the pH range of 8–11, adding Fe^{2+} forms $Fe(OH)^+$ on the sphalerite surface, which is then oxidized to a trivalent iron hydroxide complex. the addition of xanthate (X^-) generates hydrophobic $Fe(OH)_2X$ or $Fe(OH)X_2$ [108].

Depressed Effect of Flotation

Under alkaline conditions, Pb^{2+} ions give rise to the formation of a $Pb(OH)_2$ colloid within the slurry. The adsorption of $Pb(OH)_2$ onto the surface of sphalerite occurs through covalent bonding between the Pb^{2+} ions present in $Pb(OH)_2$ and the S^{2-} ions in sphalerite. Specifically, one oxygen atom from $Pb(OH)_2$ establishes a covalent bond with Zn^{2+} ions in the sphalerite structure, while another oxygen atom forms a covalent bond with hydrogen protons from neighboring $Pb(OH)_2$ molecules. Therefore, a layer of hydrophilic $Pb(OH)_2$ is covered on the surface of sphalerite, which inhibits the flotation process of sphalerite [109]. When Cu^{2+} is not present in the solution, Ca^{2+} can enhance the flotation of sphalerite under mildly alkaline conditions. However, as the pH level rises, the precipitation of $Ca(OH)_2$ on the surface of the sphalerite occurs, which subsequently hinders the flotation process [110]. When Cu^{2+} was present in the solution, Ca^{2+} or Mg^{2+} hindered the interaction between the surface of the sphalerite activated by Cu^{2+} and xanthate. This phenomenon is attributed to the precipitation of Ca and Mg on the surface of sphalerite, which forms a layer over the surface of sphalerite. Additionally, some researchers suggest that the presence of Ca^{2+} or Mg^{2+} ions may compete with Cu^{2+} ions for adsorption sites, leading to a reduction in the concentration of the activated substance Cu-S [111,112].

Marmatite is a black variety of sphalerite with a relatively high iron content (approximately 10%–26%). This is caused by the substitution of impure iron atoms in sphalerite crystals. The flotation performance of the iron sphalerite was poorer than that of ordinary sphalerite. This is because Fe^{3+} replaces the zinc atoms in the sphalerite lattice, resulting in an imbalance between the valence and charge states. This causes a portion of the Zn^{2+} to transform into Zn^{+}, reducing the hole concentration and increasing the electron density. Due to the heightened electron density, sphalerite displayed a pronounced repulsive interaction with xanthate ions, which hinders the effective adsorption of the collectors. Therefore, the floatability of marmatite is poor, and compared with ordinary sphalerite, it is not easily separated by flotation. Moreover, the presence of iron impurities has the potential to hinder the displacement rate of Cu^{2+} ions within the lattice structure of sphalerite. Even with prolonged activation, the concentration of Cu^{2+} on the sphalerite surface tends to diminish. This repulsive effect that iron impurities exert on the adsorption of xanthate results in a decreased adsorption of xanthate molecules to the surface of sphalerite, consequently compromising the flotation efficiency of the mineral [17,113,114].

In summary, Cu^{2+} serves as the most widely employed activation ion for sphalerite, effectively enhancing its flotation under both the Cu^{2+} state and the $Cu(OH)_2$ form. The presence of copper facilitates the flotation of sphalerite that has been suppressed by zinc ions. Additionally, Pb^{2+} influences the sphalerite surface through adsorption, exerting an activating effect. However, in alkaline slurry conditions, Pb^{2+} predominantly exists as lead hydroxide, which subsequently inhibits sphalerite flotation due to its adsorptive properties. The chemical adsorption of $Pb(OH)_2$ onto the sphalerite surface further hampers flotation efficiency. Although Ca^{2+} have a positive effect on the flotation of sphalerite in the presence of copper ions, its presence becomes inhibitory following the activation by Cu^{2+}. Under varying slurry conditions, Mg^{2+} and Ca^{2+} may exhibit analogous mechanisms of action. Furthermore, Fe^{2+} provides a certain degree of activation for sphalerite flotation, while Fe^{3+} does not significantly influence this process. In the case of iron sphalerite characterized by lattice defects, the presence of Fe^{3+} can alter the lattice constant, which ultimately serves to inhibit flotation.

3.2.2. The Influence of Metal Ions on the Flotation of Zinc Oxide Ore

The most industrially valuable ZnO ores are sphalerite and hemimorphite. The main flotation processes for ZnO include sulfide amine salt, sulfide xanthate, and direct fatty acid flotation [115].

Activated Effect of Flotation

Owing to the variable and mud-like nature of zinc oxide ores, conventional flotation recovery is ineffective. The incorporation of metal ions for activation can enhance the adsorption of sulfide ions and collectors on the mineral surface, subsequently leading to improved flotation efficiency.

The main reaction sites on the surface of ZnO with active metal ions are O in three different states: Zn–O, –Si-O, and –OH on the surface of minerals. Zhao et al. proposed two models. On the one hand, Pb^{2+} interacts with exposed O atoms in the slurry solution due to electrostatic interactions, forming –Zn-O-Pb and –Si-O-Pb complexes. In contrast, In the slurry solution, $Pb(OH)^{+}$ undergoes dehydration reaction with the OH^{-} component present on the surface of ZnO reaction. This process leads to the formation of –Zn-O-Pb and –Si-O-Pb complexes, which in turn generate additional active sites that facilitate the adsorption of the collector [116,117]. Zhang et al. suggest that Pb^{2+} ions are adsorbed onto the mineral surface by displacing Zn^{2+} ions present on the surface of smithsonite. Following the vulcanization process, the formation of $PbCO_3$ and $Pb_3(CO_3)_2(OH)_2$ occurs, which subsequently react with xanthate to produce hydrophobic lead xanthate [118]. Feng et al. proposed a method of ammonia surface modification to improve sulfide flotation of smithsonite, which generates zinc ammonia complexes and interacts with $Zn(OH)_2$

and $Zn(OH)_3^-$ species on the mineral surface, enriching the mineral surface with a large amount of Zn^{2+} and increasing the active sites for interaction with sulfides [119].

Similar to copper oxide sulfide flotation, metal ions positively impact the flotation of zinc oxide sulfide. Moderate amount of Pb^{2+} can increase the formation of Pb-S component on the surface of sulfide smithsonite, generating more active sites for xanthate adsorption and enhancing the reactivity of the sulfide smithsonite surface with xanthate [120]. Jia et al. suggest that Pb^{2+} binds to newly formed sulfur atoms on ZnS at the surface of sulfide hemimorphite minerals through an ion exchange mechanism, where Pb^{2+} replaces Zn^{2+}, forming Pb-S species and promoting collector adsorption [121]. Luo et al. propose that after Pb^{2+} treatment, the Pb-S substances on the surface of sulfide smithsonite may exist as Pb-S, PbS^{2-}, and PbS^{3-} complexes. The reaction between BX^- and Pb-S sites on the smithsonite surface forms $Pb(BX)_2$ species, making the surface of sphalerite more hydrophobic [122]. Contrary to the activation of a single Pb^{2+} ion, stepwise activation leads to an increase in the formation of Pb-S species on the surface of heteropolar minerals. This enhancement contributes to the adsorption of xanthate and significantly improves the hydrophobicity and floatability of mineral surfaces [123]. Further research indicates the synergistic activation effect of copper and lead ions in the flotation of sulfide hemimorphite minerals. Their research shows that the synergistic effect of copper and lead was superior to Pb^{2+} activation alone, as well as Cu^{2+} activation alone. Cu^{2+} and Pb^{2+} enter the slurry and generate various sulfides on the surface of mineral, thereby and effectively promoting the adsorption of xanthate esters [124].

Depressed Effect of Flotation

Iron ions can significantly hinder the flotation of zinc oxide ore when introduced into the slurry. The main components leading to this inhibitory effect are the iron hydroxyl component and precipitates generated on the surface of mineral due to Fe-O and hydrogen bonds [125–127]. Luo et al. investigated the inhibitory effect of Ca^{2+} on the sulfidation reaction of smithsonite. Their findings revealed that pre-treatment with Ca^{2+} significantly impedes the subsequent adsorption of HS^-. Specifically, the Ca-S bond established on the surface of smithsonite exhibits a lower strength and stability in comparison to the Zn-S bond generated following HS adsorption in the absence of prior Ca^{2+} pre-adsorption. Furthermore, after the adsorption of Ca^{2+}, HS encounters greater challenges in penetrating the hydration shell on the surface of smithsonite, which in turn results in a diminished flotation performance [128].

In summary, as shown in Figure 6 and Table 5, the products formed by active ions such as Pb^{2+} and Cu^{2+} during the sulfidation of zinc oxide ore not only increase adsorption capacity but also exhibit higher activity compared to direct sulfidation. This may be related to differences in the radius and hydration energy of metal ions. The inhibitory effect of metal ions on zinc oxide ore shows similarities, as hydroxyl ions or precipitates can hinder the surface activation process of minerals.

Table 5. Activation and inhibition effects of various metal ions in the flotation system of zinc oxide minerals.

Minerals	Metal Ions	Activating or Inhibiting Substances
Smithsonite	Cu^{2+}	CuS/ZnS
	Pb^{2+}	PbS/ZnS
	Fe^{2+}	Fe-O-OH
	Fe^{3+}	Fe-O-OH
Hemimorphite	Pb^{2+}	PbS
	Fe^{3+}	Fe-O-OH

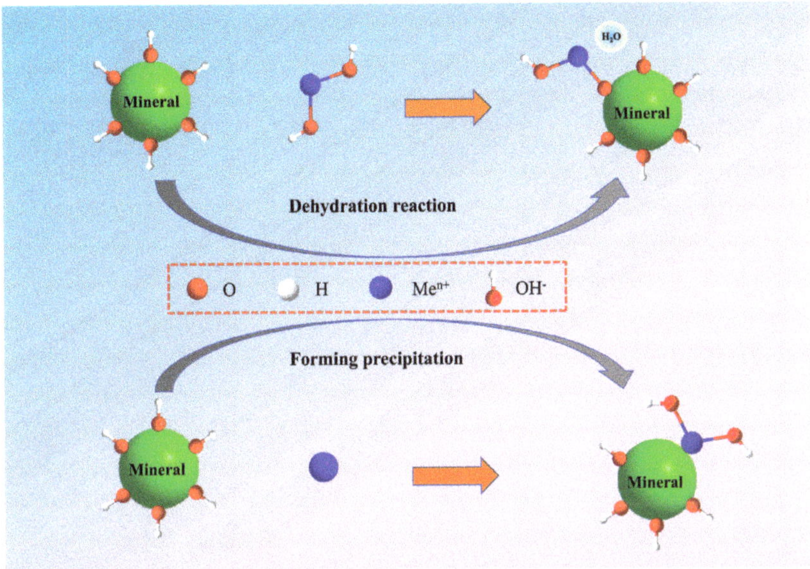

Figure 6. Model showing the formation of metal ion hydroxyl complexes on mineral surfaces and the precipitation of metal hydroxides on these surfaces.

3.3. Influence of Metal Ions on the Flotation of Lead Minerals

Lead ore can be categorized into two types: sulfide lead ore and oxide lead ore, with sulfide lead ore having significantly greater reserves and distribution than oxide lead ore [13]. Pure sulfide lead deposits are rare in nature, as lead sulfide often coexists with zinc sulfide, forming polymetallic deposits.

3.3.1. Effect of Metal Ions on the Flotation of Lead Sulfide Ore

Galena is the most important sulfide lead ore, known for its good natural floatability. Both Ca^{2+} and Mg^{2+} can depress the flotation efficiency of galena, calcium primarily adsorbs onto the surface of galena in the forms of Ca^{2+} and $Ca(OH)^+$ ions, with $Ca(OH)^+$ binding to some S atoms on the mineral surface, increasing its hydrophilicity. Under strongly alkaline conditions, Mg^{2+} generates $Mg(OH)_2$ precipitate, which covers the surface of galena and hinders collector adsorption [129,130]. Pb^{2+} enhances the depressed effect of sodium humate on galena, when the pH is below 9.4, Pb^{2+} reacts with sodium humate in solution to form a complex that inhibits collector adsorption on galena, As shown in Figure 7, at pH levels above 9.4, the Pb in the slurry primarily exists as positively charged hydroxy lead species, which are not conducive to the adsorption of sodium humate and weaken the inhibitory effect on galena [131]. Wang et al. studied the induced adsorption of pectin on the surface of Cu^{2+}-modified galena. This may be due to the formation of positively charged copper hydroxide species on the surface of galena, which facilitates the adsorption of pectin and thus inhibits galena. XPS results suggest that the pectin adsorption on the galena surface may be due to the induction of S-Cu-O and Cu-O-R component by the generated copper sulfide and hydroxide [132].

3.3.2. Effect of Metal Ions on the Flotation of Lead Oxide Ore

With a reduction in sulfide ore resources and the fact that most are characterized by poverty, fineness, and impurities, the development and utilization of oxidized lead ore has gradually become a research focus [133]. Cerussite, an important oxidized lead ore, is often recovered and utilized through sulfide xanthate flotation. Sulfurization results are usually influenced by many factors [134]. Metal ions such as Fe^{3+} or Fe^{2+} entering the flotation system will consume some Na_2S, and changing the pH of the slurry hinders the sulfurization process. The generation of hydrophilic component on the surface of mineral is also a reason for poor flotation effects [135,136]. As shown in Figure 8, The influence of low concentrations of Cu^{2+} on the flotation of cerussite is not significant, but excessive Cu^{2+} entering the slurry not only consumes the sulfurizing agent to produce a large amount of Cu-S material, but also inhibits the generation of sulfurizing products, resulting in an unsatisfactory sulfurizing effect [137].

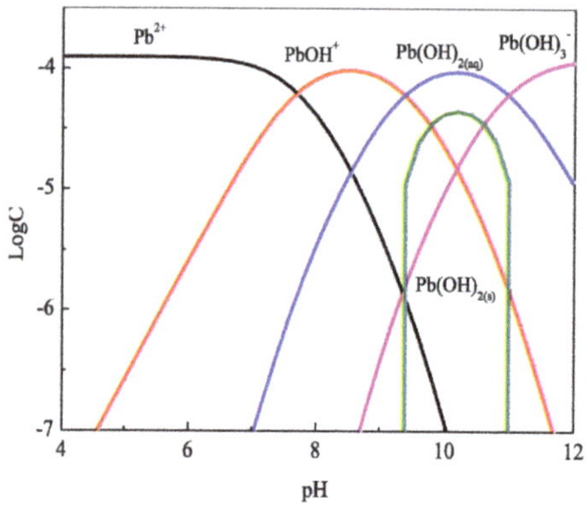

Figure 7. Species distribution diagram of the Pb^{2+}-H_2O system ($[Pb^{2+}] = 1.25 \times 10^{-4}$ mol/L) [131].

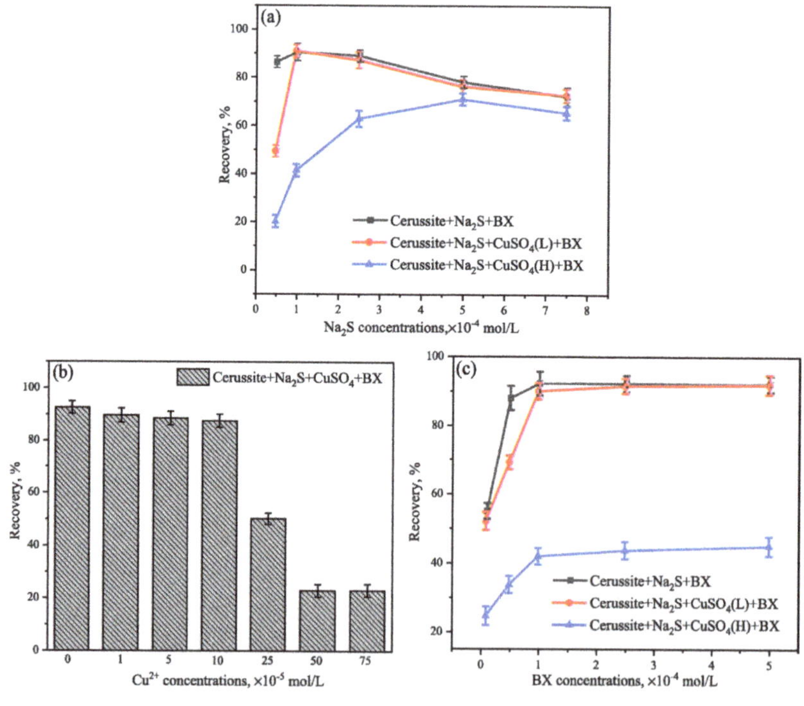

Figure 8. Flotation recoveries of sulfidized cerussite treated with various reagents: (**a**) $CuSO_4$, (**b**) Na_2S, (**c**) BX. (Na_2S: 1×10^{-3} mol/L; $CuSO_4$ (L): 1×10^{-4} mol/L, $CuSO_4$ (H): 5×10^{-4} mol/L; BX: 5×10^{-4} mol/L) [137].

4. The Influence of Metal Ions on Flotation Separation of Mixed Minerals

The action of metal ions on minerals is usually non-selective, and their introduction into the slurry system can alter the surface properties of minerals, impacting flotation separation. The abundant gangue minerals in copper, lead, and zinc ores are particularly susceptible to metal ions due to their high content [138,139]. Ca^{2+}, Mg^{2+}, Al^{3+}, and Fe^{3+} can activate quartz during flotation with anionic collectors [140–142]. This activation is thought to arise from the formation of complexes or the adsorption of metal salts of these ions onto the surface of quartz particles. Ca^{2+} has minimal impact on the flotation of talc [143], while Mg^{2+}, Cu^{2+}, and Fe^{3+} inhibit talc flotation under various conditions. As shown in Figure 9, In separating chalcopyrite from talc through flotation, Zn^{2+} forms precipitates on the talc surface under alkaline conditions, hindering the adsorption of talc and xanthate salts. This allows chalcopyrite flotation to proceed unaffected, enabling the effective separation of chalcopyrite from talc [144].

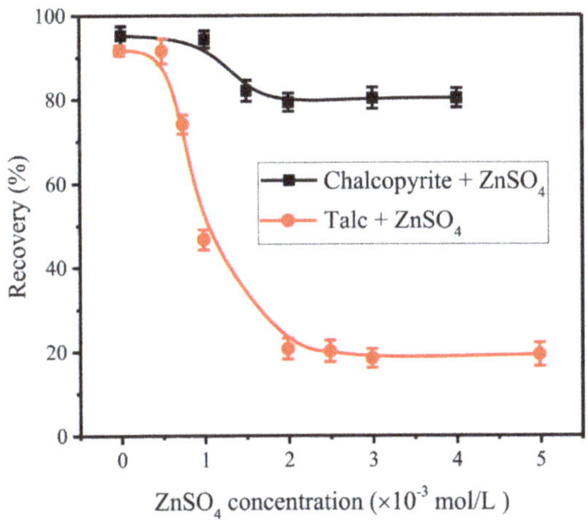

Figure 9. Microflotation results of chalcopyrite and talc as a function of the zinc sulfate concentration [144] (pH = 9, C(BX) = 2 mg/L).

Oxidized ores exhibit high solubility in slurries, and the primary reason for the suboptimal flotation separation efficiency is the activation of gangue minerals by active metal ions present in the dissolved components [145]. The copper ions dissolved in the malachite supernatant predominantly adhered to the quartz surface as Cu^{2+}, $Cu(OH)^{+}$, and $Cu(OH)_2$ species. This adsorption facilitated the binding of sulfur species and xanthate ions, which augmented the hydrophobicity of the quartz surface, complicating the flotation separation process [146]. This phenomenon has also been observed in the separation of cerussite and quartz. Dissolved Pb^{2+} acts on the surface of mineral to provide adsorption sites for sulfur ions and xanthates. As shown in Figure 10, the dissolution of calcite within the slurry leads to the substantial adsorption of Ca^{2+} ions onto the surface of the cerussite. This process covers the adsorption sites of both sulfides and collectors, consequently diminishing the hydrophobicity of the cerussite. Pb^{2+} adsorb onto the surface of calcite by bonding with the O sites in calcite, forming Pb-S species after sulfidation and promoting the adsorption of xanthates. Therefore, the disparity in hydrophobicity between the two minerals decreased, resulting in a deterioration of the flotation separation efficiency [147].

When using organic inhibitors for separation, the combined inhibitory effect of metal ions and adjusters can improve the flotation efficiency. Xie et al. formulated a way to separate smithsonite from calcite by utilizing a combination of guar gum (GG) and Al^{3+} as a mixed inhibitor. The hydrolysis products of Al^{3+} were found to be selectively adsorbed onto the surface of calcite, and GG formed Ca-O and Al-O bonds by interacting with the

Ca and Al sites on the surface of calcite, enhancing the hydrophilicity of the surface. In addition, $Al(OH)_3$ and Al_2O_3 were generated on the surface of mineral, covering the NaOL adsorption sites. The synergistic effect of these two mechanisms suppresses calcite flotation and achieves efficient separation from smithsonite [148,149].

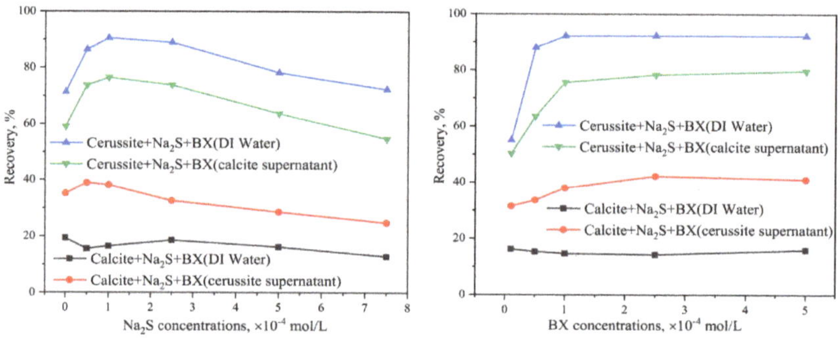

Figure 10. Flotation recoveries of cerussite and calcite under various conditions [149].

Effectively separating sulfide polymetallic minerals through flotation presents significant challenges, primarily due to the comparable surface properties of co-occurring minerals and the interactive effects of dissolved metal ions on the mineral surfaces. These metal ions have the potential to form complexes with various reagents, which can modify the differences in hydrophilicity and hydrophobicity among mineral surfaces. Consequently, this alteration can negatively impact the efficiency of flotation separation [3,150]. Under low alkaline conditions, Ca^{2+} are adsorbed on the surfaces of arsenopyrite and chalcopyrite in the form of $Ca(OH)^+$, after adsorption, compared with the calcium sulfur bonds present on chalcopyrite, $Ca(OH)^+$ species form stronger iron oxide bonds (Fe-O) and calcium sulfur bonds (Ca-S) on the surface of arsenopyrite, consequently, more $Ca(OH)^+$ is adsorbed on arsenopyrite, enhancing its hydrophilicity, while the flotation of chalcopyrite remains unaffected, leading to improved separation efficiency [151]. The Pb^{2+} released from the dissolution of galena can activate the flotation of sphalerite. DMDC (sodium dimethyl dithiocarbamate) acts as an inhibitor by forming a complex with Pb^{2+}, thereby weakening Pb^{2+} activation of sphalerite. It also combines with Zn^{2+} to compete for adsorption, reducing the collector's effect on sphalerite. Meanwhile, the floatability of galena remains unaffected, enabling flotation separation from sphalerite [152]. However, the use of CLS lignosulfonate calcium as a galena inhibitor in Pb-Cu separation has shown unsatisfactory results. Su et al. found that CLS selectivity decreased in the presence of Pb^{2+}, possibly due to the formation of charged $Pb(OH)^+$ species on the chalcopyrite surface, which promoted the physical adsorption of CLS. This increased competitive adsorption between CLS and the collector, reducing the collector's adsorption and lowering chalcopyrite's surface hydrophobicity [153]. Fenugreek gum effectively separates sphalerite from chalcopyrite, this is probably due to the fact that Zn^{2+} ions on the surface of sphalerite exhibit a greater tendency to chelate with the C-OH and O-C-O functional groups present in fenugreek gum, when compared to the interactions occurring at the Cu and Fe sites on the surface of chalcopyrite [154].

During the flotation process, complex physical and chemical reactions, including surface oxidation, ion adsorption, precipitation adsorption, and flotation agent interactions, occur on the surface of minerals. These reactions can result in the homogenization of surface chemical composition among minerals [52,155,156]. Lai et al. studied the homogenization effect between minerals in sulfide ore flotation system. As shown in Figure 11, the dissolution of chalcopyrite and galena during grinding tends to homogenize the surface chemical composition of minerals, potentially increasing the difficulty of mineral separation. Cu^{2+} and Fe^{3+} dissolve from chalcopyrite and adsorb onto galena surfaces, forming Pb-S-Cu,

Pb-O-Cu, and Pb-O-Fe component. Similarly, through the adsorption of Pb^{2+} dissolved from galena, these species also form on chalcopyrite surfaces. During the simultaneous grinding of galena and sphalerite, Pb^{2+} released from galena migrates to and adsorbs on sphalerite surfaces, while Zn^{2+} and Fe^{3+} released from sphalerite rarely adsorb onto galena. Thus, the homogeneity of grinding is influenced by the elements present in the minerals. When chalcopyrite and sphalerite are concurrently ground, the ions released from each mineral tend to migrate toward the surface of the other. Cu^{2+} from chalcopyrite more readily adsorbs onto sphalerite than Fe^{3+}, while Zn^{2+} from sphalerite poorly adsorbs onto chalcopyrite. This homogenization effect between sulfide minerals offers new insights into the challenges of flotation separation [157,158].

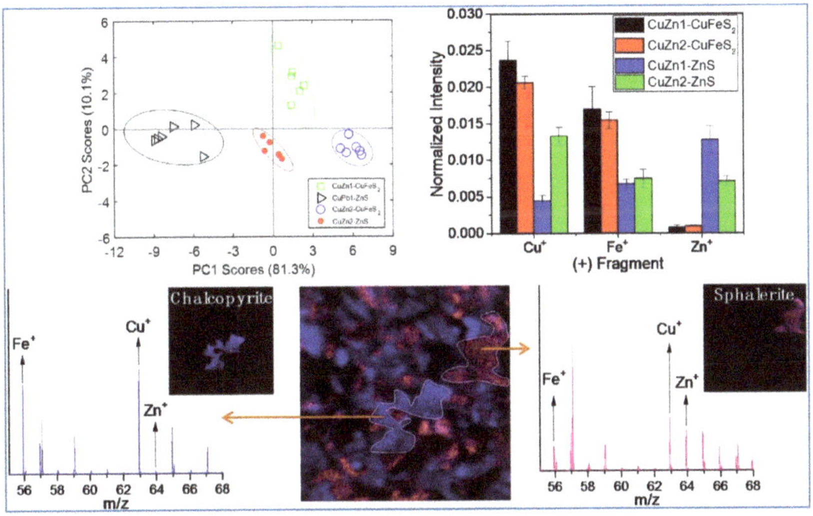

Figure 11. Homogenization phenomena of surface components of chalcopyrite and sphalerite during grinding processing [157].

When using different collectors for the flotation of galena and chalcopyrite, Cu^{2+} and Pb^{2+} exhibit similar inhibitory or activating effects on these minerals. At pH 6.5 and 9.5, Cu^{2+} primarily adsorbs onto sulfur-rich sites on the surface of galena, forming Cu(I)-S species as Pb-S-Cu-$(OH)_n$. As the pH increases, Cu^{2+} replaces Pb in the galena lattice, forming Cu_xS_y material. Meanwhile, Pb^{2+} forms $(OH)_n$-Pb-O-Fe, $(OH)_n$-Pb-S-Cu, and $(OH)_n$-Pb-O-Cu species in decreasing amounts on the surface of chalcopyrite. Whether using ethionium or butylamine as collectors, copper and lead salts corresponding to the collectors formed on the surface of Cu^{2+}-treated galena, while Cu^{2+} inhibited the production of lead salts. Similar phenomena also occur on the surface of chalcopyrite treated with Pb^{2+}. Whether or not a collector is present, the surface chemical component of Cu^{2+}-treated galena tends to shift towards that of chalcopyrite, while the surface composition of Pb^{2+}-treated chalcopyrite shifts towards that of galena. Thus, under the influence of Pb^{2+} and Cu^{2+}, the surface chemical compositions of chalcopyrite and galena tend to homogenize. In the ethyl xanthate system, $FeCl_3$ can serve as an inhibitor for chalcopyrite, thereby enhancing the separation process of chalcopyrite from galena. $FeCl_3$ promotes the oxidation and dissolution of the surface of both galena and chalcopyrite, removing Cu, S, Pb, and ethyl xanthate products from the chalcopyrite surface and forming a surface rich in hydrophilic iron (hydroxyl) oxides/hydroxides and iron sulfate. However, $FeCl_3$ has little effect on the surface composition of lead ore, thereby increasing the differences in mineral surface composition. Reducing homogenization effects can achieve selective separation [159–162].

5. Conclusions

This study delves into the adsorption properties of metal ions on mineral surfaces within slurry solutions, as well as the mechanisms of activation and inhibition of minerals during flotation processes. The principal conclusions are as follows:

During the mineral flotation process, the pulp contains a variety of metal ions, primarily originating from the flotation reagents that are added, the dissolution of associated metal minerals occurring during the crushing and grinding stages, the wear of grinding media, and the flotation water utilized. In flotation operations, enhancing the activating effects of metal ions within the pulp on mineral flotation, or mitigating the adverse impacts they may cause, can significantly enhance flotation recovery.

In the flotation process of copper, lead, and zinc minerals, the introduction of an appropriate amount of Pb^{2+} and Cu^{2+} ions can significantly enhance the hydrophobic properties of the minerals. The adsorption of these highly reactive metal ions onto mineral surfaces effectively increases the amounts of active sites available, thereby facilitating the adsorption of collectors. Furthermore, lead and copper ions play a vital role in promoting the sulfidation of oxidized ores. Conversely, the addition of iron ions may lead to the surface oxidation of the minerals, resulting in decreased mineral activity. Additionally, the introduction of Ca^{2+} and Mg^{2+} ions can result in the consumption of flotation reagents, which could hinder the efficient flotation of copper, lead, and zinc minerals.

Author Contributions: B.W.: Data Curation, Formal Analysis, Investigation, Methodology, Software, Visualization, Writing—Original Draft, and Writing—Review and Editing. Y.M.: Investigation, Methodology, Project Administration, Resources, Supervision, Validation, and Writing—Review and Editing. L.W.: Conceptualization, Data Curation, Formal Analysis, Funding Acquisition, Investigation, Methodology, Project Administration, Resources, Supervision, Validation, and Writing—Review and Editing. P.S.: Conceptualization, Data Curation, Formal Analysis, Funding Acquisition, Investigation, Methodology, Project Administration, Resources, Supervision, Validation, and Writing—Review and Editing. H.W.: Formal Analysis, Funding Acquisition, Investigation, Methodology, Project Administration, Resources, Supervision, Validation, Visualization, and Writing—Review and Editing. D.L.: Formal Analysis, Funding Acquisition, Investigation, Methodology, Project Administration, Resources, Supervision, Validation, Visualization, and Writing—Review and Editing. All authors have read and agreed to the published version of the manuscript.

Funding: This study was supported by the National Natural Science Foundation of China (Grant No. 52404285) Yunnan Major Scientific and Technological Projects (Grant No. 202202AG050015), Yunnan Provincial Science and Technology Project at Southwest United Graduate School (Grant No. 202302AO370002).

Data Availability Statement: No new data were created or analyzed in this study.

Conflicts of Interest: The authors declare no conflict of interest.

References

1. Che, Y.; Chen, W.; Liu, S.; Zeng, G.; Liu, G. Exploring the Highly Efficient Depressant Role of Fenugreek Gum on Talc in Chalcopyrite Flotation. *Miner. Eng.* **2024**, *213*, 108751. [CrossRef]
2. Shen, Z.; Feng, Q.; Wen, S.; Wang, H. Flotation Separation of Chalcopyrite from Galena Based on Selective Adsorption of Pyrogallic Acid: Experiments and MD Simulations. *Appl. Surf. Sci.* **2024**, *667*, 160398. [CrossRef]
3. Xu, X.; Zhang, X.; Zhang, Y.; Liu, X.; He, H.; Fang, J.; Shen, P.; Peng, R.; Yu, J.; Chen, X.; et al. A Study on the Mechanism of Flotation Separation between Chalcopyrite and Sphalerite Using a Novel Environmental Depressant Derived from Grape Peel Extract GPE. *Colloids Surf. A Physicochem. Eng. Asp.* **2024**, *681*, 132772. [CrossRef]
4. Yuan, J.; Li, Y.; Ding, Z.; Yu, A.; Zhang, Y.; Wen, S.; Bai, S. Influence and Mechanism of New Environmentally Friendly Depressant Carboxymethyl-β-Cyclodextrin on the Flotation Separation of Chalcopyrite and Pyrite. *Colloids Surf. A Physicochem. Eng. Asp.* **2024**, *699*, 134576. [CrossRef]
5. He, H.; Fang, J.; Qiu, Z.; Liu, D.; Xie, H.; Shen, P.; Peng, R.; Peng, L.; Qin, S.; Dong, S. Selective Galena Depression Mechanism of Tea Polyphenol in Chalcopyrite Flotation. *Miner. Eng.* **2023**, *202*, 108259. [CrossRef]
6. Cao, J.; Liao, R.; Wu, D.; Zuo, Q.; Liu, J.; Wen, S. Utilizing Phosphonyl Carboxylic Acid Copolymer as an Efficient Depressant for Flotation Separation of Chalcopyrite from Galena: Experimental and DFT Study. *Sep. Purif. Technol.* **2024**, *348*, 127725. [CrossRef]

7. Tang, H.; Deng, Z.; Tang, Y.; Tong, X.; Wei, Z. Hotspots and Trends of Sphalerite Flotation: Bibliometric Analysis. *Sep. Purif. Technol.* **2023**, *312*, 123316. [CrossRef]
8. Chen, Y.; Feng, B.; Peng, J.; Wang, Z. Selective Flotation of Galena from Sphalerite Using a Combination of $KMnO_4$ and Carboxylated Chitosan. *Appl. Surf. Sci.* **2022**, *602*, 154412. [CrossRef]
9. Zhang, S.; Feng, Q.; Wen, S.; Xian, Y.; Liu, J.; Liang, G. Flotation Separation of Chalcopyrite from Molybdenite with Sodium Thioglycolate: Mechanistic Insights from Experiments and MD Simulations. *Sep. Purif. Technol.* **2024**, *342*, 126958. [CrossRef]
10. Nowosielska, A.M.; Nikoloski, A.N.; Parsons, D.F. The Effects of NaCl Addition on the Particle-Bubble Interactions of Galena in the Presence of Xanthate. *Results Surf. Interfaces* **2024**, *14*, 100191. [CrossRef]
11. Hassanzadeh, A.; Hassas, B.V.; Kouachi, S.; Brabcova, Z.; Çelik, M.S. Effect of Bubble Size and Velocity on Collision Efficiency in Chalcopyrite Flotation. *Colloids Surf. A Physicochem. Eng. Asp.* **2016**, *498*, 258–267. [CrossRef]
12. Liu, Y.; Xu, R.; Sun, N.; Sun, W.; Liu, R.; Wang, L. Effect of Metal Ions on the Flotation Separation of Biotite from Quartz Using Mixed Anionic/Cationic Collectors. *Chem. Eng. Sci.* **2023**, *281*, 119184. [CrossRef]
13. Chen, J. The Interaction of Flotation Reagents with Metal Ions in Mineral Surfaces: A Perspective from Coordination Chemistry. *Miner. Eng.* **2021**, *171*, 107067. [CrossRef]
14. Zhang, X.; Ren, L.; Bao, S.; Zhang, Y. Effect of Iron(III) Ions on Rutile Flotation: Surface Properties and Adsorption Mechanism. *Colloids Surf. A Physicochem. Eng. Asp.* **2024**, *697*, 134343. [CrossRef]
15. Shen, Y.; Nagaraj, D.R.; Farinato, R.; Somasundaran, P.; Tong, S. Decomposition of Flotation Reagents in Solutions Containing Metal Ions. Part I: Gaseous Compounds from Xanthate Decomposition. *Miner. Eng.* **2019**, *139*, 105893. [CrossRef]
16. Yu, X.; Yu, B.; Wang, H.; Shen, P.; Liu, D. Evolution of Sulfur Species on Malachite Surfaces in the Presence of Calcium and Magnesium Ions and Implications for Xanthate Flotation. *Adv. Powder Technol.* **2024**, *35*, 104538. [CrossRef]
17. Chen, Y.; Chen, J.; Guo, J. A DFT Study on the Effect of Lattice Impurities on the Electronic Structures and Floatability of Sphalerite. *Miner. Eng.* **2010**, *23*, 1120–1130. [CrossRef]
18. Cheng, T.; Wang, T.; Shen, Z.; Ma, S.; Shi, S.; Deng, J.; Deng, J. The Depression Mechanism of Magnesium on the Flotation of Sphalerite with Different Iron Contents: DFT and Experimental Studies. *Sep. Purif. Technol.* **2025**, *353*, 128379. [CrossRef]
19. Wright, K.; Gale, J.D. A First Principles Study of the Distribution of Iron in Sphalerite. *Geochim. Cosmochim. Acta* **2010**, *74*, 3514–3520. [CrossRef]
20. Zhai, J.; Chen, P.; Wang, H.; Hu, Y.; Sun, W. Flotability Improvement of Ilmenite Using Attrition-Scrubbing as a Pretreatment Method. *Minerals* **2017**, *7*, 13. [CrossRef]
21. Parapari, P.S.; Irannajad, M.; Mehdilo, A. Modification of Ilmenite Surface Properties by Superficial Dissolution Method. *Miner. Eng.* **2016**, *92*, 160–167. [CrossRef]
22. Aarab, I.; Derqaoui, M.; El Amari, K.; Yaacoubi, A.; Abidi, A.; Etahiri, A.; Baçaoui, A. Influence of Surface Dissolution on Reagents' Adsorption on Low-Grade Phosphate Ore and Its Flotation Selectivity. *Colloids Surf. A Physicochem. Eng. Asp.* **2021**, *631*, 127700. [CrossRef]
23. Zhu, Y.G.; Zhang, G.F.; Feng, Q.M.; Yan, D.C.; Wang, W.Q. Effect of Surface Dissolution on Flotation Separation of Fine Ilmenite from Titanaugite. *Trans. Nonferrous Met. Soc. China* **2011**, *21*, 1149–1154. [CrossRef]
24. Liu, M.; Chen, D.; Hu, B.; He, P.; Chen, Y.; Zeng, H.; Zhang, C.; Zhu, J. New Insights into the Activation Mechanism of Ammonium Ions on the Malachite Sulfidization Flotation. *Miner. Eng.* **2024**, *205*, 108452. [CrossRef]
25. Wang, H.; Wen, S.; Liu, D.; Han, G.; Feng, Q. Influence of Ferrous Ions on Surface Properties of Malachite and Its Response to Sulfidization Xanthate Flotation. *Miner. Eng.* **2022**, *189*, 107880. [CrossRef]
26. Irannajad, M.; Salmani Nuri, O.; Mehdilo, A. Surface Dissolution-Assisted Mineral Flotation: A Review. *J. Environ. Chem. Eng.* **2019**, *7*, 103050. [CrossRef]
27. Bierlein, F.P.; Ashley, P.M.; Seccombe, P.K. Seccombe Origin of Hydrothermal Cu-Zn-Pb Mineralisation in the Olary Block, South Australia: Evidence from Fluid Inclusions and Sulphur Isotopes. *Precambrian Res.* **1996**, *79*, 281–305. [CrossRef]
28. Gke, A.; Bozkaya, G. Fluid Inclusion and Stable Isotope Characteristics of the Karalar (Gazipaa, Antalya) Barite-Galena Deposits, Southern Turkey. *Geol. Ore Depos.* **2008**, *50*, 145–154.
29. Wilkinson, J.J. Fluid Inclusions in Hydrothermal Ore Deposits. *Lithos* **2001**, *55*, 229–272. [CrossRef]
30. Deng, J.; Wen, S.; Xian, Y.; Liu, J.; Bai, S. New Discovery of Unavoidable Ions Source in Chalcopyrite Flotation Pulp: Fluid Inclusions. *Miner. Eng.* **2013**, *42*, 22–28. [CrossRef]
31. Bai, S.; Wen, S.; Xian, Y.; Liu, J.; Deng, J. New Source of Unavoidable Ions in Galena Flotation Pulp: Components Released from Fluid Inclusions. *Miner. Eng.* **2013**, *45*, 94–99. [CrossRef]
32. Liu, J.; Wen, S.; Wu, D.; Bai, S.; Liu, D. Determination of the Concentrations of Calcium and Magnesium Released from Fluid Inclusions of Sphalerite and Quartz. *Miner. Eng.* **2013**, *45*, 41–43. [CrossRef]
33. Finkelstein, N.P. Addendum to: The Activation of Sulphide Minerals for Flotation: A Review. *Int. J. Miner. Process.* **1999**, *55*, 283–286.
34. Wang, H.; Wen, S.; Han, G.; Feng, Q. Effect of Copper Ions on Surface Properties of $ZnSO_4$-Depressed Sphalerite and Its Response to Flotation. *Sep. Purif. Technol.* **2019**, *228*, 115756. [CrossRef]
35. Guo, B.; Peng, Y.; Espinosa-Gomez, R. Cyanide Chemistry and Its Effect on Mineral Flotation. *Miner. Eng.* **2014**, *66*, 25–32. [CrossRef]

36. Zhao, G.F.; Bo, F.E.; Zhu, D.M.; Qiu, X.H.; Gao, Z.Y.; Yan, H.S.; Lai, R.S.; Qiu, T.S. Enhanced Separation of Scheelite and Calcite by Metal-Inorganic Complex Depressant. *Trans. Nonferrous Met. Soc. China* **2024**, *34*, 643–654. [CrossRef]
37. Wan, H.; An, Y.; Luukkanen, S.; Qu, J.; Zhang, C.; Xue, J.; Wang, S.; Bu, X. New Insights into the Sphalerite Activated by Copper Sulfate in Lime Systems. *Colloids Surf. A Physicochem. Eng. Asp.* **2023**, *667*, 131397. [CrossRef]
38. Wang, H.; Wen, S.; Han, G.; Xu, L.; Feng, Q. Activation Mechanism of Lead Ions in the Flotation of Sphalerite Depressed with Zinc Sulfate. *Miner. Eng.* **2020**, *146*, 106132. [CrossRef]
39. Wills, B.A.; Finch, J. *Wills' Mineral Processing Technology: An Introduction to the Practical Aspects of Ore Treatment and Mineral Recovery*; Butterworth-Heinemann: Oxford, UK, 2015.
40. Carlson, J.J.; Kawatra, S.K. Effects of CO_2 on the Zeta Potential of Hematite. *Int. J. Miner. Process.* **2011**, *98*, 8–14. [CrossRef]
41. Corin, K.C.; Reddy, A.; Miyen, L.; Wiese, J.G.; Harris, P.J. The Effect of Ionic Strength of Plant Water on Valuable Mineral and Gangue Recovery in a Platinum Bearing Ore from the Merensky Reef. *Miner. Eng.* **2011**, *24*, 131–137. [CrossRef]
42. Levay, G.; Smart, R.S.C.; Skinner, W.M. The Impact of Water Quality on Flotation Performance. *J. South. Afr. Inst. Min. Metall.* **2001**, *101*, 69–75.
43. Peters, N.E.; Meybeck, M. Water Quality Degradation Effects on Freshwater Availability: Impacts of Human Activities. *Water Int.* **2000**, *25*, 185–193. [CrossRef]
44. Wang, B.; Peng, Y. The Effect of Saline Water on Mineral Flotation—A Critical Review. *Miner. Eng.* **2014**, *66–68*. [CrossRef]
45. Li, Y.; Xie, S.; Zhao, Y.; Xia, L.; Li, H.; Song, S. The Life Cycle of Water Used in Flotation: A Review. *Min. Metall. Explor.* **2018**, *36*, 385–397. [CrossRef]
46. Song, N.; Sun, C.; Yin, W.; Yao, J.; Yang, B. Exploration on Flotation Behavior of Galena in Seawater and Related Mechanism. *Physicochem. Probl. Miner. Process.* **2022**, *58*, 151524. [CrossRef]
47. Byrne, P.; Wood, P.J.; Reid, I. The Impairment of River Systems by Metal Mine Contamination: A Review Including Remediation Options. *Crit. Rev. Environ. Sci. Technol.* **2012**, *42*, 2017–2077. [CrossRef]
48. Song, N.; Yin, W.; Yao, J. New Insights on the Role of Seawater in Sulfide Ore Flotation—A Review. *Powder Technol.* **2024**, *431*, 119128. [CrossRef]
49. Mu Yongjun, Y.P. The Effect of Saline Water on Copper Activation of Pyrite in Chalcopyrite Flotation. *Miner. Eng.* **2019**, *131*, 336–341.
50. Dos Santos, I.D.; Oliveira, J.F. Utilization of Humic Acid as a Depressant for Hematite in the Reverse Flotation of Iron Ore. *Miner. Eng.* **2007**, *20*, 1003–1007. [CrossRef]
51. Lai, R.W.M.; Stone, L.C.; Rimmasch, B.E. Effect of Humus Organics on the Flotation Recovery of Molybdenite. *Int. J. Miner. Process.* **1984**, *12*, 163–172. [CrossRef]
52. Bruckard, W.J.; Sparrow, G.J.; Woodcock, J.T. A Review of the Effects of the Grinding Environment on the Flotation of Copper Sulphides. *Int. J. Miner. Process.* **2011**, *100*, 1–13. [CrossRef]
53. Gonçalves, K.L.C.; Andrade, V.L.L.; Peres, A.E.C. The Effect of Grinding Conditions on the Flotation of a Sulphide Copper Ore. *Miner. Eng.* **2003**, *16*, 1213–1216. [CrossRef]
54. Hao, J.; Liu, J.; Lai, H.; Liao, R.; Gao, H.; Bai, X. A New Application of Cu^{2+} on Differential Modification to Promote Copper-Molybdenum Separation with a Novel Chalcopyrite Depressant Amidinothiourea. *Sep. Purif. Technol.* **2025**, *353*, 128282. [CrossRef]
55. Gül, A.; Yüce, A.E.; Sirkeci, A.A.; Özer, M. Use of Non-Toxic Depressants in the Selective Flotation of Copper-Lead-Zinc Ores. *Can. Metall. Q.* **2013**, *47*, 111–118. [CrossRef]
56. Hu, Y.; Wu, M.; Liu, R.; Sun, W. A Review on the Electrochemistry of Galena Flotation. *Miner. Eng.* **2020**, *150*, 106272. [CrossRef]
57. Matsuoka, H.; Mitsuhashi, K.; Kawata, M.; Tokoro, C. Derivation of Flotation Kinetic Model for Activated and Depressed Copper Sulfide Minerals. *Minerals* **2020**, *10*, 1027. [CrossRef]
58. Suyantara, G.P.W.; Hirajima, T.; Miki, H.; Sasaki, K.; Kuroiwa, S.; Aoki, Y. Effect of H_2O_2 and Potassium Amyl Xanthate on Separation of Enargite and Tennantite from Chalcopyrite and Bornite Using Flotation. *Miner. Eng.* **2020**, *152*, 106371. [CrossRef]
59. Bilal, M.; Ito, M.; Koike, K.; Hornn, V.; Hiroyoshi, N. Effects of Coarse Chalcopyrite on Flotation Behavior of Fine Chalcopyrite. *Miner. Eng.* **2021**, *163*, 106776. [CrossRef]
60. Cao, Z.; Chen, X.; Peng, Y. The Role of Sodium Sulfide in the Flotation of Pyrite Depressed in Chalcopyrite Flotation. *Miner. Eng.* **2018**, *119*, 93–98. [CrossRef]
61. Dong, S.; He, H.; Qiu, Z.; Peng, L.; Qin, S.; Fang, J.; Liu, D.; Shen, P. An Eco-Friendly Depressant Derived from Grape Seed Extract (GSE) as Depressant for Flotation Separation of Pyrite from Chalcopyrite under Low-Alkalinity Conditions. *Miner. Eng.* **2024**, *213*, 108758. [CrossRef]
62. Dai, Z.; Zheng, Y.; Guo, Z.; Peng, J.; Jian, S.; Wang, Z. Selective Depression Mechanism of Polyaspartic Acid and Calcium Oxide on Arsenopyrite after Copper Ions Activation and Its Effect on Flotation Separation Performance. *J. Hazard. Mater.* **2024**, *473*, 134689. [CrossRef] [PubMed]
63. Sun, X.; Yu, J.; Zhang, W.; Li, P.; Li, Y.; Han, Y. Flotation Separation of Chalcopyrite from Pyrite and Sphalerite Using a Novel Collector: An Experimental and Mechanism Investigation. *Appl. Surf. Sci.* **2024**, *661*, 160059. [CrossRef]
64. Liu, Y.; Wei, Z.; Xue, C. Selective Depression of Pb^{2+}-Activated Sphalerite by Potassium Ferricyanide in Pb-Zn Sulfides Flotation Separation. *Miner. Eng.* **2022**, *182*, 107558. [CrossRef]

65. Deng, J.S.; Wen, S.M.; Liu, J.; Wu, D.D.; Feng, Q.C. Adsorption and Activation of Copper Ions on Chalcopyrite Surfaces: A New Viewpoint of Self-Activation. *Trans. Nonferrous Met. Soc. China* **2014**, *24*, 3955–3963. [CrossRef]
66. Wang, C.; Liu, R.; Zhai, Q.; Wu, M.; Jing, N.; Xie, F.; Sun, W. Influence of Calcium and Ferric Ions on the Depression of Chalcopyrite by CMC: Flotation Performance and Adsorption Mechanism Study. *Miner. Eng.* **2022**, *184*, 107667. [CrossRef]
67. Dong, J.; Liu, Q.; Yu, L.; Subhonqulov, S.H. The Interaction Mechanism of Fe^{3+} and NH_4^+ on Chalcopyrite Surface and Its Response to Flotation Separation of Chalcopyrite from Arsenopyrite. *Sep. Purif. Technol.* **2021**, *256*, 117778. [CrossRef]
68. Hao, J.; Liu, J.; Gao, H.; Yu, Y.; Bai, X. Auxiliary Depression of Copper Ion—Strengthening the Depression of Sodium Trithiocarbonate on Chalcopyrite. *Miner. Eng.* **2023**, *204*, 108395. [CrossRef]
69. Northey, S.; Mohr, S.; Mudd, G.M.; Weng, Z.; Giurco, D. Modelling Future Copper Ore Grade Decline Based on a Detailed Assessment of Copper Resources and Mining. *Resour. Conserv. Recycl.* **2014**, *83*, 190–201. [CrossRef]
70. Zhang, L.; Cai, Z.; Yang, J.; Yuan, Z.; Chen, Y. The Future of Copper in China—A Perspective Based on Analysis of Copper Flows and Stocks. *Sci. Total Environ.* **2015**, *536*, 142–149. [CrossRef]
71. Dai, P.; Xue, Z.; Li, X.; Wei, Z.; Zeng, J.; Chen, L. Magnetic Property of Malachite for High-Gradient Magnetic Separation. *Miner. Eng.* **2024**, *213*, 108734. [CrossRef]
72. Cai, J.; Yu, X.; Su, C.; Pei, B.; Shen, P.; Liu, D. Improved Azurite Sulfurization with Ethylenediamine Modification and Its Effects on Flotation Behaviors: Insight from Copper Vacancy. *Miner. Eng.* **2024**, *209*, 108636. [CrossRef]
73. Jiang, D.; Lan, J.; Zhao, W.; Zhang, Z.; Lan, Y. Activation of Chrysocolla Flotation by Organic Chelating Agents. *RSC Adv.* **2017**, *7*, 35608–35612. [CrossRef]
74. Jia, X.D.; Song, K.W.; Cai, J.P.; Chao, S.U.; Xu, X.H.; Shen, P.L.; Liu, D.W. Effect of Oxygen and Sodium Sulfide on Flotation of Cuprite and Its Modification Mechanism. *Trans. Nonferrous Met. Soc. China* **2023**, *33*, 1233–1243. [CrossRef]
75. Zhang, Q.; Wen, S.; Feng, Q.; Wang, H. Enhanced Sulfidization of Azurite Surfaces by Ammonium Phosphate and Its Effect on Flotation. *Int. J. Miner. Metall. Mater.* **2022**, *29*, 1150–1160. [CrossRef]
76. Kongolo, K.; Kipoka, M.; Minanga, K.; Mpoyo, M. Improving the Efficiency of Oxide Copper–Cobalt Ores Flotation by Combination of Sulphidisers. *Miner. Eng.* **2003**, *16*, 1023–1026. [CrossRef]
77. Li, Z.; Rao, F.; Guo, B.; Zuo, W.; Song, S.; López-Valdivieso, A. Effects of Calcium Ions on Malachite Flotation with Octyl Hydroxamate. *Miner. Eng.* **2019**, *141*, 105854. [CrossRef]
78. Feng, Q.; Zhao, W.; Wen, S.; Cao, Q. Copper Sulfide Species Formed on Malachite Surfaces in Relation to Flotation. *J. Ind. Eng. Chem.* **2017**, *48*, 125–132. [CrossRef]
79. Feng, Q.; Zhao, W.; Wen, S. Surface Modification of Malachite with Ethanediamine and Its Effect on Sulfidization Flotation. *Appl. Surf. Sci.* **2018**, *436*, 823–831. [CrossRef]
80. Feng, Q.; Zhao, W.; Wen, S. Ammonia Modification for Enhancing Adsorption of Sulfide Species onto Malachite Surfaces and Implications for Flotation. *J. Alloys Compd.* **2018**, *744*, 301–309. [CrossRef]
81. Jin, J.; Miller, J.D.; Dang, L.X.; Wick, C.D. Effect of Cu^{2+} Activation on Interfacial Water Structure at the Sphalerite Surface as Studied by Molecular Dynamics Simulation. *Int. J. Miner. Process.* **2015**, *145*, 66–76. [CrossRef]
82. Wang, H.; Wen, S.; Han, G.; He, Y.; Feng, Q. Adsorption Behavior and Mechanism of Copper Ions in the Sulfidization Flotation of Malachite. *Int. J. Min. Sci. Technol.* **2022**, *32*, 897–906. [CrossRef]
83. Cao, S.; Yin, W.; Yang, B.; Zhu, Z.; Sun, H.; Sheng, Q.; Chen, K. Insights into the Influence of Temperature on the Adsorption Behavior of Sodium Oleate and Its Response to Flotation of Quartz. *Int. J. Min. Sci. Technol.* **2022**, *32*, 399–409. [CrossRef]
84. Wang, H.; Wen, S.; Han, G.; Feng, Q. Modification of Malachite Surfaces with Lead Ions and Its Contribution to the Sulfidization Flotation. *Appl. Surf. Sci.* **2021**, *550*, 149350. [CrossRef]
85. Zhang, Q.; Wen, S.; Feng, Q.; Liu, J. Surface Modification of Azurite with Lead Ions and Its Effects on the Adsorption of Sulfide Ions and Xanthate Species. *Appl. Surf. Sci.* **2021**, *543*, 148795. [CrossRef]
86. Zhang, Q.; Wang, Y.; Feng, Q.; Wen, S.; Zhou, Y.; Nie, W.; Liu, J. Identification of Sulfidization Products Formed on Azurite Surfaces and Its Correlations with Xanthate Adsorption and Flotation. *Appl. Surf. Sci.* **2020**, *511*, 145594. [CrossRef]
87. Yang, W.; Feng, Q.; Han, G. A Novel Activation Approach for Promoting Chrysocolla Flotation: Performance and Mechanism. *Appl. Surf. Sci.* **2023**, *640*, 158426. [CrossRef]
88. Wang, H.; Wen, S.; Han, G.; Feng, Q. Adsorption Characteristics of Pb(II) Species on the Sulfidized Malachite Surface and Its Response to Flotation. *Sep. Purif. Technol.* **2021**, *264*, 118440. [CrossRef]
89. Zhang, Q.; Wen, S.; Feng, Q.; Liu, Y. Activation Mechanism of Lead Ions in the Flotation of Sulfidized Azurite with Xanthate as Collector. *Miner. Eng.* **2021**, *163*, 106809. [CrossRef]
90. Wang, H.; Wen, S.; Han, G.; Feng, Q. Interaction Mechanism of Copper Ions with the Surface of Sulfidized Malachite and Its Response to Flotation. *Colloids Surf. A Physicochem. Eng. Asp.* **2022**, *647*, 129127. [CrossRef]
91. Lu, W.; Wen, S.; Liu, D.; Wang, H.; Feng, Q. A Novel Method for Improving Sulfidization Xanthate Flotation of Malachite: Copper–Ammonium Synergistic Activation. *Appl. Surf. Sci.* **2023**, *618*, 156660. [CrossRef]
92. Cai, J.; Su, C.; Ma, Y.; Yu, X.; Peng, R.; Li, J.; Zhang, X.; Fang, J.; Shen, P.; Liu, D. Role of Ammonium Sulfate in Sulfurization Flotation of Azurite: Inhibiting the Formation of Copper Sulfide Colloid and Its Mechanism. *Int. J. Min. Sci. Technol.* **2022**, *32*, 575–584. [CrossRef]
93. Zhang, Q.; Wen, S.; Zuo, Q.; Feng, Q. Density Functional Theory Calculation of Catalytic Sulfidization on the Azurite (0 1 1) Surface by Ammonium Salts and Its Effect on Flotation. *Sep. Purif. Technol.* **2023**, *309*, 122993. [CrossRef]

94. Yang, X.; Liu, S.; Liu, G.; Zhong, H. A DFT Study on the Structure–Reactivity Relationship of Aliphatic Oxime Derivatives as Copper Chelating Agents and Malachite Flotation Collectors. *J. Ind. Eng. Chem.* **2016**, *46*, 404–415. [CrossRef]
95. Barbaro, M.; Urbina, R.H.; Cozza, C.; Fuerstenau, D.; Marabini, A. Flotation of Oxidized Minerals of Copper Using a New Synthetic Chelating Reagent as Collector. *Int. J. Miner. Process.* **1997**, *50*, 275–287. [CrossRef]
96. Feng, Q.; Yang, W.; Wen, S.; Wang, H.; Zhao, W.; Han, G. Flotation of Copper Oxide Minerals: A Review. *Int. J. Min. Sci. Technol.* **2022**, *32*, 1351–1364. [CrossRef]
97. Xiaojun, X.; Bangrui, L. Mechanism of Fine Malachite Floatation by Activation with Triethanolamine. *Nonferrous Met.* **1993**, *1*, 28–33.
98. Yu, X.; Shen, P.; Yin, Z.; Wang, L.; Wang, H.; Liu, D. Surface Modification of Malachite Using DMTD and Its Effect on Xanthate Adsorption. *Colloids Surf. A Physicochem. Eng. Asp.* **2023**, *679*, 132560. [CrossRef]
99. Li, F.; Zhong, H.; Xu, H.; Jia, H.; Liu, G. Flotation Behavior and Adsorption Mechanism of α-Hydroxyoctyl Phosphinic Acid to Malachite. *Miner. Eng.* **2015**, *71*, 188–193. [CrossRef]
100. Wang, H.; Wen, S.; Liu, D.; Han, G.; Feng, Q. Surface Characteristic and Sulfidization-Xanthate Flotation Behaviours of Malachite as Influenced by Ferric Ions. *Colloids Surf. A Physicochem. Eng. Asp.* **2023**, *668*, 131469. [CrossRef]
101. Zhang, H.; Zhou, Q.; Lin, S.; Zhang, C.; Sun, W.; Chen, D.; Wang, R.; Liu, M.; Chen, J. Surface Modification of Malachite with Tetraamminecopper (II) and Its Effect on Sulfidation Flotation. *Miner. Eng.* **2022**, *189*, 107882. [CrossRef]
102. Buckley, A.N.; Woods, R.; Wouterlood, H.J. An XPS Investigation of the Surface of Natural Sphalerites under Flotation-Related Conditions. *Int. J. Miner. Process.* **1989**, *26*, 29–49. [CrossRef]
103. Gu, G.; Wang, D.; Liu, R. Electrochemical Mechanisms on Cupric Sulphate Activating Sphalerite. *J. Cent. South Univ. Technol.* **1999**, *30*, 374–377.
104. Ralston, O.C.; King, C.R.; Tartaron, F.X. Copper Sulfate as Flotation Activator for Sphalerite. *Trans. AimE* **1930**, *87*, 389–400.
105. Liu, J.; Zeng, Y.; Luo, D.; Wen, S.; Wang, Y. *Ab Initio* Molecule Dynamic Simulation of $Cu(OH)_2$ Interaction with Sphalerite (1 1 0) Surface. *Miner. Eng.* **2018**, *122*, 176–178. [CrossRef]
106. Steele, H.M.; Wright, K.; Hillier, I.H. A Quantum-Mechanical Study of the (110) Surface of Sphalerite (ZnS) and Its Interaction with Pb^{2+} Species. *Phys. Chem. Miner.* **2003**, *30*, 69–75. [CrossRef]
107. Pattrick, R.A.D.; Charnock, J.M.; England, K.E.R.; Mosselmans, J.F.W.; Wright, K. Lead Sorption on the Surface of ZnS with Relevance to Flotation: A Fluorescence Reflexafs Study. *Miner. Eng.* **1998**, *11*, 1025–1033. [CrossRef]
108. Zhang, Q.; Rao, S.R.; Finch, J.A. Flotation of Sphalerite in the Presence of Iron Ions. *Colloids Surf.* **1992**, *66*, 81–89. [CrossRef]
109. Sarvaramini, A.; Larachi, F.; Hart, B. Collector Attachment to Lead-Activated Sphalerite—Experiments and DFT Study on PH and Solvent Effects. *Appl. Surf. Sci.* **2016**, *367*, 459–472. [CrossRef]
110. Dávila-Pulido, G.I.; Uribe-Salas, A.; Álvarez-Silva, M.; López-Saucedo, F. The Role of Calcium in Xanthate Adsorption onto Sphalerite. *Miner. Eng.* **2015**, *71*, 113–119. [CrossRef]
111. Ejtemaei, M.; Plackowski, C.; Nguyen, A.V. The Effect of Calcium, Magnesium, and Sulphate Ions on the Surface Properties of Copper Activated Sphalerite. *Miner. Eng.* **2016**, *89*, 42–51. [CrossRef]
112. Lascelles, D.; Finch, J.A.; Sui, C. Depressant Action of Ca and Mg on Flotation of Cu Activated Sphalerite. *Can. Metall. Q.* **2003**, *42*, 133–140. [CrossRef]
113. Chen, Y.; Chen, J.; Lan, L.; Yang, M. The Influence of the Impurities on the Flotation Behaviors of Synthetic ZnS. *Miner. Eng.* **2012**, *27–28*, 65–71. [CrossRef]
114. Liu, J.; Wang, Y.; Luo, D.; Chen, L.; Deng, J. Comparative Study on the Copper Activation and Xanthate Adsorption on Sphalerite and Marmatite Surfaces. *Appl. Surf. Sci.* **2018**, *439*, 263–271. [CrossRef]
115. Zhang, Y.; Zhao, W.; Han, G.; Feng, Q. A Novel Activation Method for Regulating Surface Characteristics and Flotation Performance of Smithsonite: XPS, SEM-EDS, AFM, ToF-SIMS and FT-IR Studies. *Sep. Purif. Technol.* **2025**, *352*, 128276. [CrossRef]
116. Zhao, W.; Liu, D.; Wen, S.; Feng, Q. Surface Modification of Hemimorphite with Lead Ions and Its Effect on Flotation and Oleate Adsorption. *Appl. Surf. Sci.* **2019**, *483*, 849–858. [CrossRef]
117. Feng, Q.; Wang, M.; Zhang, G.; Zhao, W.; Han, G. Enhanced Adsorption of Sulfide and Xanthate on Smithsonite Surfaces by Lead Activation and Implications for Flotation Intensification. *Sep. Purif. Technol.* **2023**, *307*, 122772. [CrossRef]
118. Zhang, S.; Wen, S.; Xian, Y.; Liang, G.; Li, M. Pb Ion Pre-Modification Enhances the Sulfidization and Floatability of Smithsonite. *Miner. Eng.* **2021**, *170*, 107003. [CrossRef]
119. Feng, Q.; Wen, S.; Bai, X.; Chang, W.; Cui, C.; Zhao, W. Surface Modification of Smithsonite with Ammonia to Enhance the Formation of Sulfidization Products and Its Response to Flotation. *Miner. Eng.* **2019**, *137*, 1–9. [CrossRef]
120. Wang, M.; Zhao, W.; Han, G.; Feng, Q. Utilization of Lead Ions to Improve Surface Hydrophobicity and Flotation Recovery of Sulfidized Smithsonite. *Colloids Surf. A Physicochem. Eng. Asp.* **2023**, *663*, 131126. [CrossRef]
121. Jia, K.; Feng, Q.; Zhang, G.; Shi, Q.; Chang, Z. Understanding the Roles of Na_2S and Pb(II)in the Flotation of Hemimorphite. *Miner. Eng.* **2017**, *111*, 167–173. [CrossRef]
122. Luo, B.; Nie, W.; Dong, J. Effect of Lead Ions on the Sulfidization Flotation of Smithsonite Using Sodium Butyl Xanthate as a Collector. *Miner. Eng.* **2022**, *185*, 107710. [CrossRef]
123. Yi, Y.; Li, P.; Zhang, G.; Feng, Q.; Han, G. Stepwise Activation of Hemimorphite Surfaces with Lead Ions and Its Contribution to Sulfidization Flotation. *Sep. Purif. Technol.* **2022**, *299*, 121679. [CrossRef]

124. Feng, Q.; Zhang, G.; Zhang, Q.; Zhao, W. Synergistic Activation of Sulfidized Hemimorphite with Copper-Lead Species for Improving Surface Hydrophobicity and Floatability. *Sep. Purif. Technol.* **2024**, *332*, 125854. [CrossRef]
125. Feng, Q.; Zhao, G.; Zhang, G.; Zhao, W.; Han, G. Degradation Mechanism of Surface Hydrophobicity by Ferrous Ions in the Sulfidization Flotation System of Smithsonite. *Colloids Surf. A Physicochem. Eng. Asp.* **2022**, *648*, 129119. [CrossRef]
126. Li, P.; Zhang, G.; Zhao, W.; Han, G.; Feng, Q. Interaction Mechanism of Fe^{3+} with Smithsonite Surfaces and Its Response to Flotation Performance. *Sep. Purif. Technol.* **2022**, *291*, 121001. [CrossRef]
127. Luo, Y.; Ou, L.; Chen, J.; Zhang, G.; Xia, Y.; Zhu, B.; Zhou, H. Atomic-Level Insights into the Modification Mechanism of Fe (III) Ion on Smithsonite (1 0 1) Surface from DFT Calculation. *Adv. Powder Technol.* **2022**, *33*, 103695. [CrossRef]
128. Luo, Y.; Ou, L.; Zhang, G.; Chen, J.; Luo, Y.; Zhou, H.; Yang, H.; Yin, C. Unveiling the Role of Ca Ion in the Sulfidation of Smithsonite: A Density Functional Theory Study. *J. Mol. Liq.* **2022**, *367*, 120485. [CrossRef]
129. Ikumapayi, F.; Makitalo, M.; Johansson, B.; Rao, K.H. Recycling of Process Water in Sulphide Flotation: Effect of Calcium and Sulphate Ions on Flotation of Galena. *Miner. Eng.* **2012**, *39*, 77–88. [CrossRef]
130. Ningbo, S.; Wanzhong, Y.; Jin, Y. Effect of Seawater on Flotation Separation of Galena from Sphalerite. *Adv. Powder Technol.* **2024**, *35*, 104333. [CrossRef]
131. Cui, Y.; Jiao, F.; Qin, W.; Yang, C.; Wang, X.; Li, W.; Zhang, Z.; Li, X. Effects of Pb(II) on the Flotation Behaviour of Galena with Sodium Humate as Depressant. *Miner. Eng.* **2022**, *188*, 107854. [CrossRef]
132. Wang, C.; Liu, R.; Wu, M.; Zhai, Q.; Sun, W.; Jing, N.; Xie, F. Induced Adsorption of Pectin on Copper-Ion-Modified Galena Surfaces: Flotation and Adsorption Mechanism. *Miner. Eng.* **2021**, *173*, 107229. [CrossRef]
133. Tang, X.; Chen, J.; Chen, Y. A Density Functional Based Tight Binding (DFTB+) Study on the Sulfidization-Xanthate Flotation Mechanism of Cerussite. *Appl. Surf. Sci.* **2023**, *612*, 155677. [CrossRef]
134. Liu, R.; Liu, D.; Li, J.; Liu, S.; Liu, Z.; Gao, L.; Jia, X.; Ao, S. Improved Understanding of the Sulfidization Mechanism in Cerussite Flotation: An XPS, ToF-SIMS and FESEM Investigation. *Colloids Surf. A Physicochem. Eng. Asp.* **2020**, *595*, 124508. [CrossRef]
135. Shen, Z.; Wen, S.; Miao, Y.; Wang, H.; Feng, Q. Degradation Mechanism of Surface Sulfidization and Hydrophobicity by Fe(III) in the Flotation of Cerussite. *J. Mol. Liq.* **2023**, *385*, 122406. [CrossRef]
136. Shen, Z.; Wen, S.; Miao, Y.; Wang, H.; Feng, Q. Interaction of Fe^{2+} with Cerussite Surfaces and Its Depression Effect on Adsorption of Sulfidizing Agent and Collector. *Miner. Eng.* **2023**, *203*, 108349. [CrossRef]
137. Zhang, Q.; Wen, S.; Feng, Q. Effect of $CuSO_4$ on Sulfidized Cerussite Surface Properties and Its Response to the Flotation Behavior. *Appl. Surf. Sci.* **2022**, *589*, 152956. [CrossRef]
138. Ejtemaeia, M.; Irannajada, M.; Gharabaghi, M. Role of Dissolved Mineral Species in Selective Flotation of Smithsonite from Quartz Using Oleate as Collector. *Int. J. Miner. Process.* **2012**, *114–117*, 40–47. [CrossRef]
139. Hao, J.; Liu, J.; Yu, Y.; Gao, H.; Qin, X.; Bai, X. Depressants for Separation of Chalcopyrite and Molybdenite: Review and Prospects. *Miner. Eng.* **2023**, *201*, 108209. [CrossRef]
140. Zhou, Z.A.; Xu, Z.; Masliyah, J.H.; Czarnecki, J. Coagulation of Bitumen with Fine Silica in Model Systems. *Colloids Surf. A Physicochem. Eng. Asp.* **1999**, *148*, 199–211. [CrossRef]
141. Chen, Y.; Chen, Y.; Guo, X.; Liao, Z.; Huang, J. The Role of Phosphate in Inhibiting the Activation of Quartz Flotation Induced by Mg^{2+}. *J. Mol. Liq.* **2024**, *398*, 124278. [CrossRef]
142. Zhang, J.; Wang, W.; Liu, J.; Huang, Y.; Feng, Q. Fe(III) as an Activator for the Flotation of Spodumene, Albite, and Quartz Minerals. *Miner. Eng.* **2014**, *61*, 16–22.
143. Jin, S.; Shi, Q.; Feng, Q.; Zhang, G.; Chang, Z. The Role of Calcium and Carbonate Ions in the Separation of Pyrite and Talc. *Miner. Eng.* **2018**, *119*, 205–211. [CrossRef]
144. Jin, S.; Zhang, P.; Ou, L. Study on the Depression Mechanism of Zinc Sulfate on Talc in Chalcopyrite Flotation. *Colloids Surf. A Physicochem. Eng. Asp.* **2021**, *619*, 126474. [CrossRef]
145. Fornasiero, D.; Ralston, J. Cu(II) and Ni(II) Activation in the Flotation of Quartz, Lizardite and Chlorite. *Int. J. Miner. Process.* **2005**, *76*, 75–81. [CrossRef]
146. Shen, Z.; Tao, J.; Wen, S.; Wang, H.; Zhang, Q.; Feng, Q. Surface Characteristics and Flotation Performance of Quartz in the Presence of Dissolved Components of Malachite. *Colloids Surf. A: Physicochem. Eng. Asp.* **2023**, *656*, 130497. [CrossRef]
147. Zhang, Q.; Wen, S.; Nie, W.; Feng, Q. Effect of Dissolved Species of Cerussite on Quartz Flotation in Sulfidization Xanthate System. *J. Mol. Liq.* **2022**, *356*, 119055. [CrossRef]
148. Xie, X.; Li, B.; Xie, R.; Tong, X.; Li, Y.; Zhang, S.; Li, J.; Song, Q. Al^{3+} Enhanced the Depressant of Guar Gum on the Flotation Separation of Smithsonite from Calcite. *J. Mol. Liq.* **2022**, *368*, 120759. [CrossRef]
149. Zhang, Q.; Wen, S.; Zhang, S.; Feng, Q. Surface Chemistry of Dissolved Species of Cerussite and Calcite and Its Effect on Flotation Performance. *Colloids Surf. A Physicochem. Eng. Asp.* **2022**, *646*, 128945. [CrossRef]
150. Yu, L.; Liu, Q.; Li, S.; Deng, J.; Lai, H. Depression Mechanism Involving Fe^{3+} during Arsenopyrite Flotation. *Sep. Purif. Technol.* **2019**, *222*, 109–116. [CrossRef]
151. Luo, Y.; Xia, Y.; Zhou, H.; Yin, C.; Yang, H.; Chen, J.; Ou, L. Effect of Calcium Ions on Surface Properties of Chalcopyrite and Arsenopyrite and Its Response to Flotation Separation under Low-Alkalinity Conditions. *Appl. Surf. Sci.* **2022**, *602*, 154191. [CrossRef]
152. Cui, Y.F.; Fen, J.I.; Qin, W.Q.; Dong, L.Y.; Xu, W.A. Synergistic Depression Mechanism of Zinc Sulfate and Sodium Dimethyl Dithiocarbamate on Sphalerite in Pb–Zn Flotation System. *Trans. Nonferrous Met. Soc. China* **2020**, *30*, 2547–2555. [CrossRef]

153. Su, C.; Pei, B.; Ibrahim, A.M.; Jia, X.; Cai, J.; Shen, P.; Liu, D. Effect of Lead Ions on Accidental Adsorption of Calcium Lignosulphonate on Chalcopyrite Surface. *J. Mol. Liq.* **2023**, *378*, 121564. [CrossRef]
154. Wang, C.; Liu, R.; Wu, M.; Zhai, Q.; Luo, Y.; Jing, N.; Xie, F.; Sun, W. Selective Separation of Chalcopyrite from Sphalerite with a Novel Depressant Fenugreek Gum: Flotation and Adsorption Mechanism. *Miner. Eng.* **2022**, *184*, 107653. [CrossRef]
155. Grano, S. The Critical Importance of the Grinding Environment on Fine Particle Recovery in Flotation. *Miner. Eng.* **2009**, *22*, 386–394. [CrossRef]
156. Wang, X.H.; Xie, Y. The Effect of Grinding Media and Environment on the Surface Properties and Flotation Behaviour of Sulfide Minerals. *Miner. Process. Extr. Metall. Rev.* **1990**, *7*, 49–79. [CrossRef]
157. Lai, H.; Deng, J.; Wen, S.; Wu, D. Homogenization Phenomena of Surface Components of Chalcopyrite and Sphalerite during Grinding Processing. *Colloids Surf. A Physicochem. Eng. Asp.* **2019**, *578*, 123601. [CrossRef]
158. Lai, H.; Liu, Q.; Deng, J.; Wen, S. Using ToF-SIMS to Study Metal Ions Transfer between Chalcopyrite and Galena during Grinding. *Adv. Powder Technol.* **2020**, *31*, 2650–2657. [CrossRef]
159. Lai, H.; Deng, J.; Liu, Q.; Wen, S.; Song, Q. Surface Chemistry Investigation of Froth Flotation Products of Lead-Zinc Sulfide Ore Using ToF-SIMS and Multivariate Analysis. *Sep. Purif. Technol.* **2021**, *254*, 117655. [CrossRef]
160. Lai, H.; Liu, Q.; Deng, J.; Wen, S.; Liu, Z. Surface Chemistry Study of Cu-Pb Sulfide Ore Using ToF-SIMS and Multivariate Analysis. *Appl. Surf. Sci.* **2020**, *518*, 146270. [CrossRef]
161. Lai, H.; Shen, P.; Liu, R.; Liu, D.; Wen, S. Cryo-ToF-SIMS Study of Sodium Isobutyl Xanthate Adsorption on Sulfide Minerals. *Miner. Eng.* **2022**, *186*, 107723. [CrossRef]
162. Wang, H.; Xie, Z.; Cai, J.; Peng, R.; Xiang, Y.; Shen, P.; Liu, D. Interaction of $NH_3 \cdot H_2O$–2,5-Dimercapto-1,3,4-Thiadiazole System with Azurite Surface and Its Role in Promoting Activation Flotation. *Appl. Surf. Sci.* **2024**, *656*, 159704. [CrossRef]

minerals

Article

Flotation Separation of Smithsonite from Calcite Using Cupferron as a Collector

Qingqing Wang, Lei Sun *, Yang Cao, Xin Wang, Yi Qiao, Guobin Liu, Meitao Xiang and Wei Sun

School of Minerals Processing and Bioengineering, Central South University, Changsha 410083, China; wangqingqing@csu.edu.cn (Q.W.); caoyang@csu.edu.cn (Y.C.); xin.w_csu@csu.edu.cn (X.W.); qiaoyi@csu.edu.cn (Y.Q.); liuguobin@csu.edu.cn (G.L.); xiang-csu@csu.edu.cn (M.X.); sunmenghu@csu.edu.cn (W.S.)

* Correspondence: sunlei@csu.edu.cn; Tel.: +86-731-88830482

Abstract: The flotation separation of smithsonite and calcite is difficult due to their similar surface properties. In this study, cupferron was applied as a collector to realize the separation of smithsonite and calcite. Micro-flotation experiment results indicated that smithsonite and calcite express different floatability after treatment with cupferron. The maximum recovery difference was 63%, from a cupferron concentration of 2×10^{-4} mol/L at pH 8. Based on a series of tests, including an adsorption test, Fourier-transform infrared (FTIR), zeta potential and X-ray photoelectron spectroscopy (XPS), the selective collection mechanism of cupferron was studied. It was found that the cupferron was more easily adsorbed on the surface of smithsonite and the reaction was violent. The adsorption capacity of the cupferron on the surface of smithsonite was higher than that of calcite, and the surface potential shift was greater. The cupferron chelated with the exposed Zn sites on the smithsonite surface to form a N-O-Zn ring structure. This special chelate structure caused the smithsonite surface to be more hydrophobic, which confirmed that the cupferron can selectively collect smithsonite instead of calcite.

Keywords: smithsonite; cupferron; flotation; collector

1. Introduction

As an important nonferrous metal, zinc is widely used in electrical, mechanical, and metallurgical applications due to its excellent abrasion and corrosion resistance [1]. In the last few decades, most of the world's zinc has been produced from zinc sulfide (ZnS) concentrates [2,3]. With the gradual depletion of zinc sulfide, the proportion of zinc extracted from zinc oxide concentrates has been increasing. Hence, the flotation of smithsonite, one of the major minerals in zinc oxide ore, has attracted considerable attention [4,5].

At present, the most common flotation method for smithsonite is sulfidation-amine or sulfidation-xanthate [4], which is widely used in industry. The method applies to most zinc oxide ores [6–9]. For this method, the surface of smithsonite is converted to a sulfide surface by using sulfiding agents. Many studies have already explored the mechanism of sulfidation in order to improve the recovery of smithsonite [10]. Surface sulfidation flotation has been extensively studied, including screening sulfurizing reagents, characterizing sulfide films on smithsonite surfaces and promoting the sulfidation process of smithsonite [11–15]. However, this method has drawbacks, such as environmental pollution and excessive reagent consumption. Sulfidation flotation is not the only way to recover smithsonite; fatty acids or chelating agents can also be used to recover smithsonite directly.

Fatty acids are usually used in direct flotation for zinc oxide ore if the gangue is clay or silica minerals [16]. New collectors development has also been reported, such as with lauryl phosphate, N,N′-dilauroyl ethylenediamine dipropionate and benzohydroxamic acid [17–19]. The central problem for direct flotation of oxidized zinc ores is poor selectivity, which is needed to separate gangue minerals into concentrate. Calcite is a common gangue mineral and is frequently associated with smithsonite in zinc oxide deposits. Both smithsonite and calcite belong to the group of carbonate minerals, and the properties of their surfaces are remarkably similar [20]. For this reason, the semi-solubility of the mineral results in adverse effects from calcite on the smithsonite surface. Some approaches for separating smithsonite from calcite pay more attention to calcite depressants, such as starch, water glass, calcium lignosulphonate or carboxymethyl cellulose [21–23], but the effectiveness of these depressants is facing more challenges as the average calcite content in worldwide raw zinc oxide ore has begun to increase.

Chelating agents are a class of specific complexing agents that can interact with certain metal ions to form a ring structure [24]. The formation of water-insoluble metal chelates on the surfaces of various minerals enhances the hydrophobicity of these mineral surfaces. Cupferron, a well-known analytical reagent, is widely used in the detection of trace metal ions. It is also considered to be an effective collector of oxide and sulfide minerals. Some investigations have reported that cupferron can form monodentate or bidentate complexes with active metal sites on mineral surfaces, such as chalcopyrite, uranium oxide and malachite [25–27]. Cupferron is used as the ligand in ion flotation to separate gallium and aluminum, demonstrating its high selectivity [28]. Bottei has studied the thermal stability and infrared spectra of different cupferron complexes, and this work showed that zinc complexes have higher stability than other complexes [29]. The cupferron has good chelating properties for zinc, it could be used as a potential collector for the flotation of smithsonite.

In this study, with the aim of providing an effective chelating agent for smithsonite flotation, we evaluated cupferron as a chelate collector in the separation of smithsonite from calcite. Different flotation behaviors of smithsonite and calcite were observed through micro-flotation and adsorption experiments. Furthermore, the adsorption mechanism of cupferron on the smithsonite surface was illustrated with Fourier-transform infrared spectra (FTIR), zeta potential and X-ray photoelectron spectroscopy (XPS).

2. Materials and Methods

2.1. Samples and Reagents

High-purity samples of smithsonite and calcite minerals were obtained from Yunnan Province, China. Samples were crushed in a laboratory jaw crusher and ground in a porcelain ball mill to reduce the particle size. Then, standard sieves were used to obtain samples of several different particle sizes. The ground −74 + 38 μm was used for micro-flotation and adsorption experiments. The −5 μm fraction samples were utilized for FTIR and XPS analysis. The XRD spectra of the smithsonite and calcite particles are shown in Figure 1. Table 1 shows that the purities of the smithsonite and calcite were 96.02% and 97.68%, respectively.

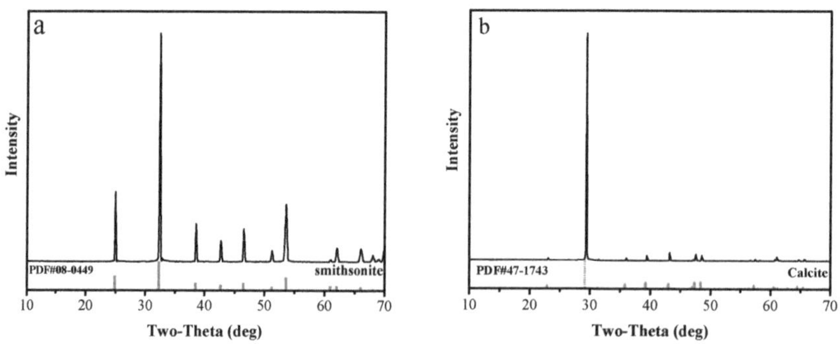

Figure 1. XRD of (**a**) smithsonite samples and (**b**) calcite samples.

Table 1. The chemical composition of smithsonite and calcite samples (wt.%).

Sample	Zn	Ca	Fe	Al	Si
Smithsonite	49.93	—	0.77	0.0056	0.0041
Calcite	—	39.07	0.016	0.0049	0.0010

Analytical grade (98%) cupferron and terpineol used in this study were purchased from Macklin Biochemical Company (Shanghai, China). The molecular structure of cupferron is shown in Figure 2 and its chemical name is N-nitroso-N-phenylhydroxylamine ammonium salt. Cupferron and terpineol were used as collectors and frothers, respectively. Sodium hydroxide (NaOH) and hydrochloric acid (HCl) were used as pH regulators. Deionized water (DI) was used throughout the experimental work.

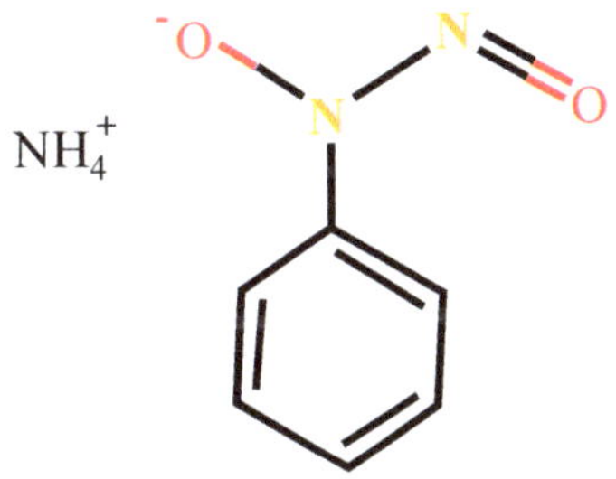

Figure 2. The molecular structure formula of cupferron.

2.2. Micro-Flotation Experiments

The single-mineral and mixed-mineral micro-flotation tests were performed in an XFG flotation machine with a fixed speed of 1650 rpm. For the mixed-mineral micro-flotation tests, the sample was fully mixed with a ratio of 1:1 smithsonite and calcite. For each test, 2 g of mineral samples and 38 mL of DI water were added to the flotation cell. The pH value was adjusted by adding HCl or NaOH until the required value was attained. Cupferron was added as a collector to the pulp and agitated for 5 min. Subsequently, terpineol was added to the pulp for 1 min more of agitation. Finally, the flotation stage was performed for 5 min. Both floated and non-floated samples were filtered, dried, and weighed to calculate the flotation recovery. To ensure the reliability of the flotation results, each test was repeated three times under the same conditions. The standard deviation was calculated and presented as an error bar in the flotation results.

2.3. Adsorption Experiments

For the adsorption experiments, the adsorption quantity of reagents on the mineral's surface was detected by using an ultraviolet-visible spectrum analyzer (UV2600, Shimadzu Corporation, Kyoto, Japan). For this analysis, 2 g mineral samples were used in each test, and the samples were treated in the same way as in the flotation test. After the agent reacted with the mineral, the solution was centrifuged by the TG16-WS machine (Thermo Fisher Scientific Inc., Waltham, MA, USA) at 9000 r/min for 15 min, and then the centrifuged solution was sent to the ultraviolet-visible spectrophotometer for residual concentration detection. The specific surface area of the minerals was measured by N_2 adsorption on a specific surface analyzer (TriStar II 3flex, micro, Norcross, GA, USA). The smithsonite surface area was 0.52 m^2/g, and the calcite surface area was 0.46 m^2/g. The adsorption capacity was calculated as follows:

$$\Gamma = \frac{(C_0 - C)V}{M \times S}$$

Γ represents the adsorption quantity of cupferron on the mineral surface. C_0 and C are the initial and supernatant cupferron concentrations. V is the volume of the solution, and M is the weight of the mineral sample. S is the specific surface area of the smithsonite or calcite sample.

2.4. FTIR Spectroscopy Measurements

The smithsonite and calcite samples before and after the adsorption were measured by an infrared spectrometer (Nicolet 6700, Thermo Fisher, Waltham, MA, USA). We tested the FTIR spectra in the range of 400 cm^{-1}~4000 cm^{-1} wavenumber. In a beaker, 2 g of minerals (<5 μm) and 38 mL of DI water were magnetically stirred. The specific concentration of cupferron was then added to the beaker after adjusting the pH of the slurry. After stirring for 30 min, the samples were filtered and placed in a low-temperature vacuum-drying oven. Finally, 100 mg of KBr and 5 mg of the mineral samples were mixed for infrared analysis.

2.5. Zeta Potential Measurement

The zeta potential measurements of smithsonite and calcite were performed on the Malvern Zetasizer instrument (Nano-ZS900, Malvern Inc., Malvern, Worcestershire, UK). The 50 mg sample below 5 μm was dispersed in 1×10^{-3} mol/L KCl electrolyte solution with magnetic stirring at room temperature. The pH was adjusted to the expected value with HCl or NaOH. According to the test conditions, a certain amount of cupferron was selectively added to the slurry. After the cupferron fully reacted with the mineral, it was left for 10 min and 2 mL of supernatant was removed to measure zeta potential. For each sample, the average of three independent measurements was taken as the final value.

2.6. XPS

The XPS measurement was performed on the ESCALAB 250Xi (Thermo Fisher, Waltham, MA, USA) using monochromatic Al Kα radiation. The mineral sample (−5 μm) was treated with cupferron concentrations of 2×10^{-4} mol/L, the blank sample was treated with DI water, and the stirring time was 30 min. After the reaction, the samples were filtered and vacuum-dried for XPS analysis. The C 1s peak of 284.8 eV was used to calibrate the spectra.

3. Results and Discussion

3.1. Micro-Flotation Tests

3.1.1. Single-Mineral Micro-Flotation Test

The pH condition test started at the neutral pH, considering the self-solubility of minerals. As shown in Figure 3, the pH value of pulp had different influences on the floatability of smithsonite and calcite. In the presence of cupferron, the recovery of smithsonite slowly increased in the range of pH 6 to 8, and the recovery of smithsonite reached its maximum of 79.08% at pH value 8. When the pH value exceeded 8.0, the recovery of smithsonite began to decline sharply, and the recovery was only 18.23% at pH 10. Compared with the maximum recovery rate, this is a decrease of 60.85%. Over the entire pH range, the recovery of calcite was low and remained stable in a small range around 15%. The chemical equilibrium of the smithsonite solution has been discussed in previous research, which showed that the species Zn^{2+} predominates until pH 8.5 [30]. Under this condition, cupferron can better chelate with smithsonite surface. Therefore, the optimum condition for separating smithsonite and calcite is pH 8.

Figure 4 presents the effects of the cupferron concentration on the flotation performance of smithsonite and calcite at pH 8. As is shown in this figure, the recovery rates of smithsonite and calcite were different when the concentration of cupferron increased. The recovery rate of smithsonite increased rapidly when the concentration of cupferron was in the range of 0 to 2×10^{-4} mol/L. The recovery of smithsonite achieved 80.67% at a cupferron concentration of 2×10^{-4} mol/L. As the concentration of cupferron exceeded 2×10^{-4} mol/L, the smithsonite recovery fluctuated around 80%. In contrast, the recovery of calcite was about 15% in the entire cupferron concentration range.

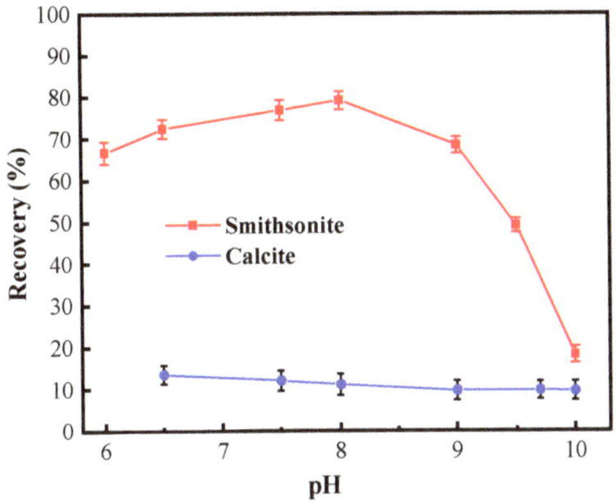

Figure 3. Flotation recovery of smithsonite and calcite as a function of pH using 2×10^{-4} mol/L cupferron.

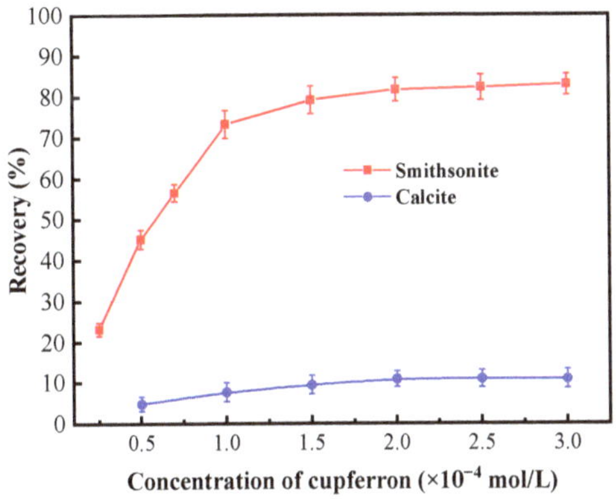

Figure 4. Flotation recovery of smithsonite and calcite as a function of the concentration of cupferron at pH 7.5~8.0.

3.1.2. Mixed-Minerals Micro-Flotation Test

The micro-flotation test of mixed minerals was performed further to verify the specific collecting performance of cupferron on smithsonite. The pulp pH was adjusted to 8, and the concentration of cupferron was 2×10^{-4} mol/L. The test results are shown in Table 2.

Table 2. The results of mixed-minerals micro-flotation test/%.

Products	Yield	Grade		Recovery	
		Zn	Ca	Zn	Ca
Concentrates	61.36	31.65	8.13	78.53	25.70
Tailings	38.64	13.74	37.34	21.47	74.30
Total	100	24.73	19.42	100	100

Table 2 shows that the recovery rate of smithsonite could reach 78.53% and that the recovery of calcite was 25.70% in the mixed-mineral flotation test. Compared with the single-mineral flotation test, the recovery of calcite was slightly improved. Previous studies have shown that mineral interactions and dissolved ions in mixed-mineral systems interfere with mineral recovery. Zn ions can disrupt the interactions between reagents and calcite by forming insoluble surface precipitates [31]. Furthermore, the presence of Ca^{2+} prevents the adsorption of sulfur ions onto smithsonite surface due to hydrolysis and precipitation, thus affecting the floatability of mineral [10]. Despite the interference of these influencing factors, on the whole, cupferron can be used to separate smithsonite from calcite.

3.2. Adsorption Experiments

From Figure 5, it is obvious that a fixed absorption peak centered at 282.5 nm can be found by measuring the absorbance of cupferron at several different concentrations. The fitted curve indicates a linear relationship between absorbance and concentration in the range of 0.5 to 3 × 10^{-4} mol/L. Figure 6 illustrates that the adsorption capacities of smithsonite and calcite to cupferron are quite different, and the adsorption amount of smithsonite was significantly higher than that of calcite. The surface adsorption amount of smithsonite increased rapidly at a concentration of 0 to 5 × 10^{-4} mol/L and reached a maximum of 17.36 × 10^{-6} mol/m^2 at a concentration of 5 × 10^{-4} mol/L, tending to stabilize as the cupferron increased. For calcite, the adsorption capacity of calcite increased slowly and was lower than 6 × 10^{-6} mol/m^2 over the whole cupferron concentration range. It can be concluded that cupferron was adsorbed on the surface of smithsonite but hardly on the surface of calcite. These results were consistent with those of the micro-flotation tests.

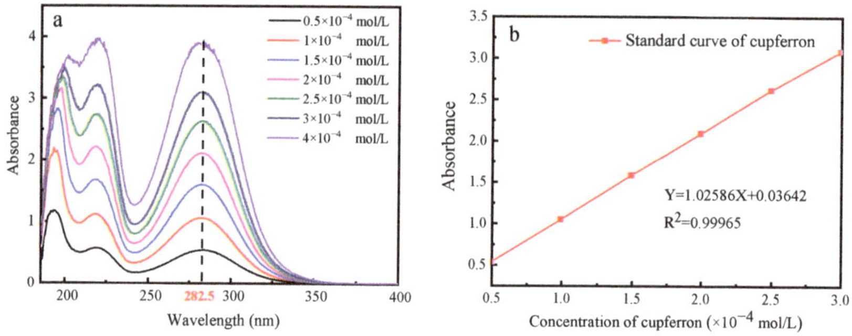

Figure 5. The UV absorption spectra of cupferron as a function of cupferron concentration (**a**) and the standard curve of cupferron with different concentrations (**b**).

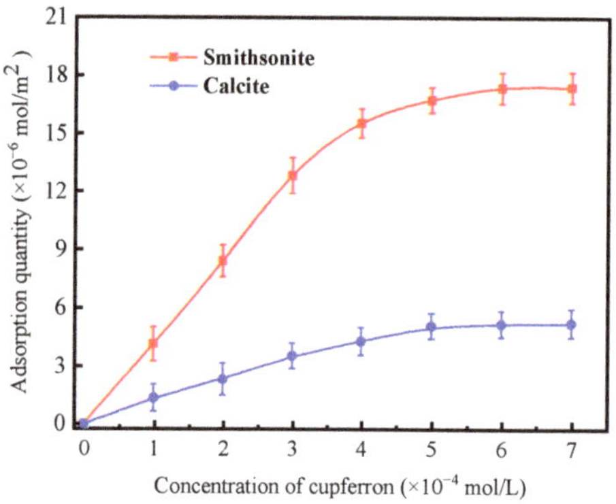

Figure 6. Adsorption quantity of cupferron on smithsonite and calcite surfaces as a function of concentration at pH 8.

3.3. FTIR Analysis

The infrared spectra of cupferron are shown in Figure 7. The strong absorption band around 3060 cm^{-1} can be attributed to the stretching mode of the NH_4 group [29], and the aromatic ring vibrations were consistent with monosubstituted benzene ring vibrations. The bands at 1592, 1457 and 1421 cm^{-1} were due to C=C stretching modes, and the four bands at 1157, 1103, 1054 and 1017 cm^{-1} in the spectra of cupferron corresponded to the C-H in-plane bending modes of the benzene ring. The only remaining C-H vibrational modes were assigned to a band at 687 cm^{-1}. The C-N stretching vibration should come around 1332 cm^{-1}, and the bands at 1266 and 1220 cm^{-1} were ascribed to the N=O stretching modes. Finally, the bands around 1332 cm^{-1} and 907 cm^{-1} of cupferron were assigned to the C-N and the N-O stretching modes [32].

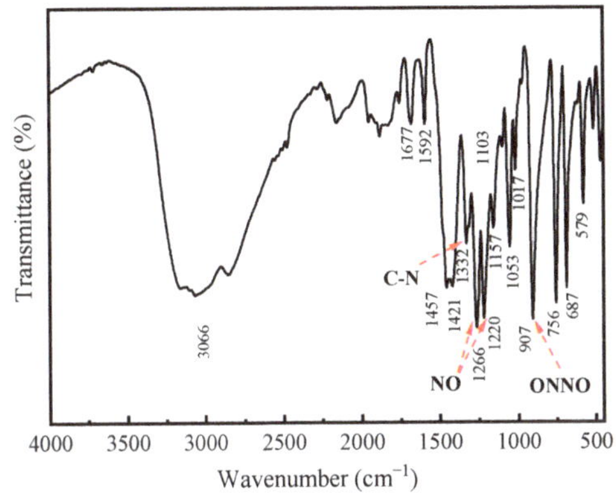

Figure 7. FTIR spectra of cupferron.

The FTIR spectra of smithsonite and cupferron-treated smithsonite are illustrated in Figure 8. The IR spectra of smithsonite showed an intense broad band at 1426 cm^{-1} that could have been from the CO_3^{2-} stretching modes [33,34], and the peaks at 869 and 743 cm^{-1} were assigned to CO_3^{2-} bending modes. Compared to the FTIR spectra of cupferron-treated smithsonite, the band at 426 cm^{-1} was ascribed to the Zn-O modes that shifted to 418 cm^{-1} [35,36]. Except for the unique groups of smithsonite, we could also see three new bands at 1089, 1049 and 923 cm^{-1}. The bands at 923 cm^{-1} could be ascribed to the single-bonded N-O absorption. Bottei and Schneggenburger have shown that the N-O group shifts at a high frequency in metal complexes [29]. The C-H of the phenyl group should be observed at around 1000 to 1200 cm^{-1}, and the bonds were indeed found at 1089 and 1049 cm^{-1} in our experiment. We concluded that cupferron adsorbed on the surface of smithsonite, and the N-O group of cupferron was found in the FTIR spectra of cupferron-treated smithsonite. This provided evidence for the chemical adsorption of cupferron on the surface of smithsonite.

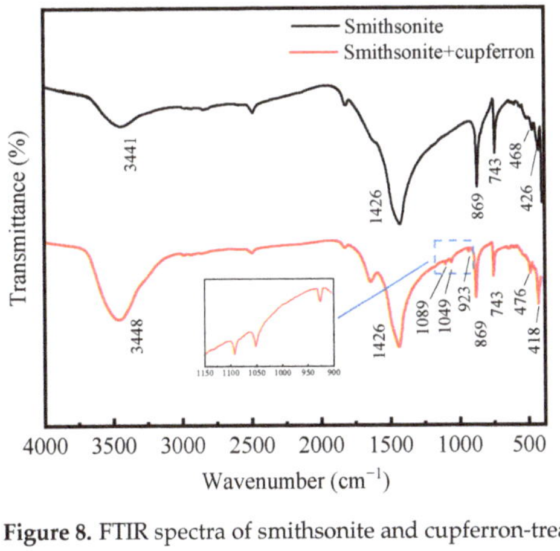

Figure 8. FTIR spectra of smithsonite and cupferron-treated smithsonite.

Figure 9 shows the FTIR spectra of calcite and cupferron-treated calcite. As with smithsonite, the spectra of calcite were characterized by the vibration mode of the CO_3^{2-} molecule. There were three main peaks in the spectrum located at 1428, 876 and 712 cm^{-1} that were ascribed to CO_3^{2-} vibration mode. Compared with the FTIR spectra of calcite, no new IR bands were observed for cupferron-treated calcite, especially in the region of 800 to 1300 cm^{-1}. These results indicated that no chemisorption occurred on the calcite surface.

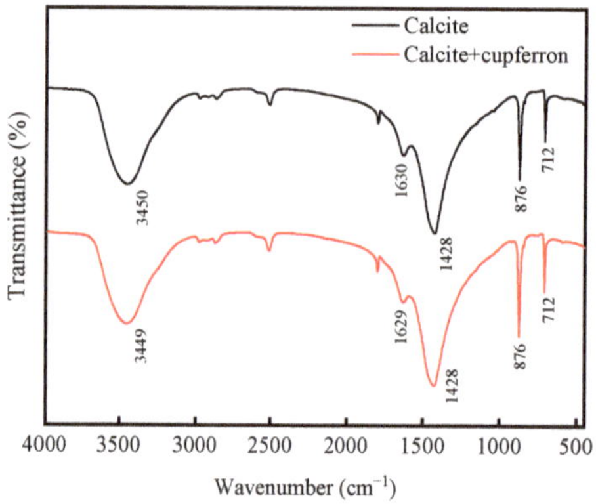

Figure 9. FTIR spectra of calcite and cupferron-treated calcite.

3.4. Zeta Potential

When minerals react with reagents, the surface potential of the minerals will change. The zeta potential of mineral surface is measured to reveal the adsorption behavior of reagents on mineral surface. The zeta potential of calcite and smithsonite before and after cupferron treatment is shown in Figure 10. The results showed that the isoelectric points of calcite and smithsonite were at pH 7.2 and 8.0, which agreed with previous reports [37,38]. With the addition of cupferron, the zeta potential of smithsonite and calcite were both shifted negatively, indicating a certain adsorption of cupferron on the two minerals' surfaces. It is more obvious that the zeta potential of smithsonite decreases more than that of calcite. There may be some differences in the adsorption methods of smithsonite and calcite. The adsorption of cupferron by smithsonite is a strong chemical adsorption, mainly due to the chemical adsorption of nitrosohydroxylamine ions on the surface of smithsonite, which leads to a high zeta potential shift, while a small amount of cupferron is physically adsorbed on the surface of calcite. This adsorption is weak and unstable, so it will cause a slight shift in the zeta potential of calcite.

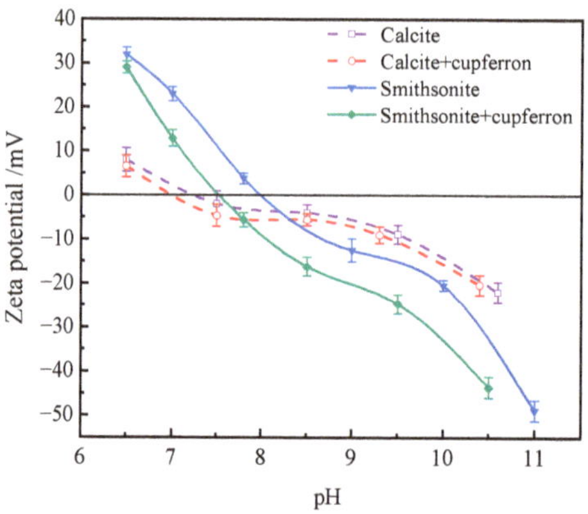

Figure 10. Influences of cupferron addition on the zeta potential of smithsonite and calcite.

3.5. XPS Analysis

The XPS survey spectra of smithsonite and cupferron-treated smithsonite are shown in Figure 11. In the XPS of smithsonite, there were no other peaks observed except for the peculiar Zn 2p, O 1s and C 1s peaks, demonstrating the purity of the sample. New peaks at N 1s were observed in the XPS survey spectra of cupferron-treated smithsonite, indicating that the cupferron adsorbed on the mineral surface. However, as shown in Figure 12, although specific peaks of Ca 2p, O 1s and C 1s were observed in both calcite and cupferron-treated calcite, no obvious peaks appeared around 400 eV related to N 1s in the XPS of cupferron-treated calcite. It can be concluded that the cupferron tended to adsorb on the surface of smithsonite but not calcite.

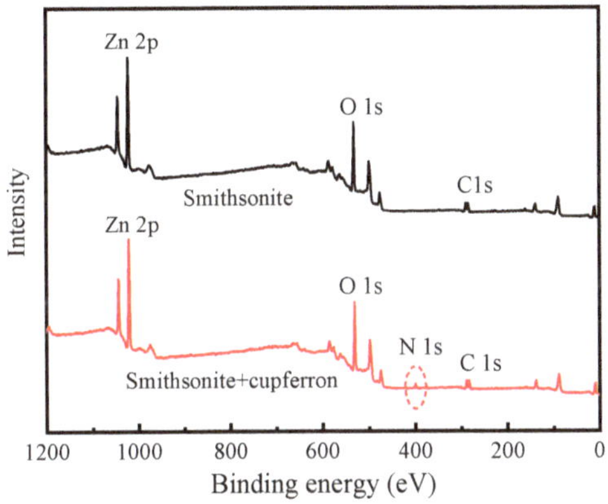

Figure 11. Survey scan XPS spectra of smithsonite and cupferron-treated smithsonite.

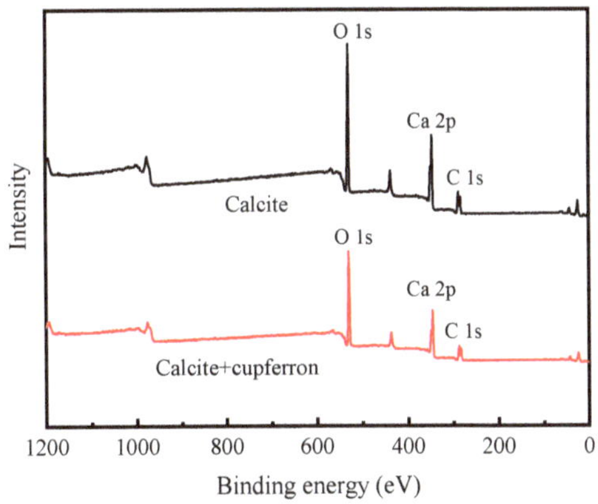

Figure 12. Survey scan XPS spectra of calcite and cupferron-treated calcite.

Figure 13 exhibits the high-resolution spectra of Zn 2p. The peaks positioned at 1022.19 eV and 1045.19 eV corresponded to the binding energies of Zn 2p3/2 and Zn 2p1/2. After the smithsonite was treated with cupferron, the two peaks of Zn 2p3/2 and Zn 2p1/2 shifted 0.26 eV towards the lower binding energy. Figure 14 shows the high-resolution O1s XPS spectra of smithsonite treated with cupferron. For bare smithsonite samples, this peak can be divided into two peaks centered at the binding energies of 531.75 eV and 532.40 eV, which are assigned to Zn hydroxide and Zn carbonate on the surface of the smithsonite [39,40]. After treatment with cupferron, this peak was divided into three peaks centered at the binding energies of 531.67 eV, 532.10 eV and 532.80 eV. The peaks were offset at Zn carbonate and Zn hydroxide, and there was an extra peak at 532.10 eV that could be ascribed to the N-O group [41].

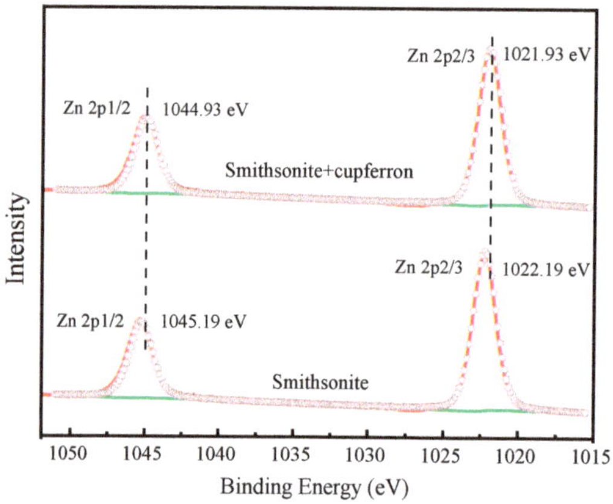

Figure 13. Zn 2p XPS analysis of smithsonite and cupferron-treated smithsonite.

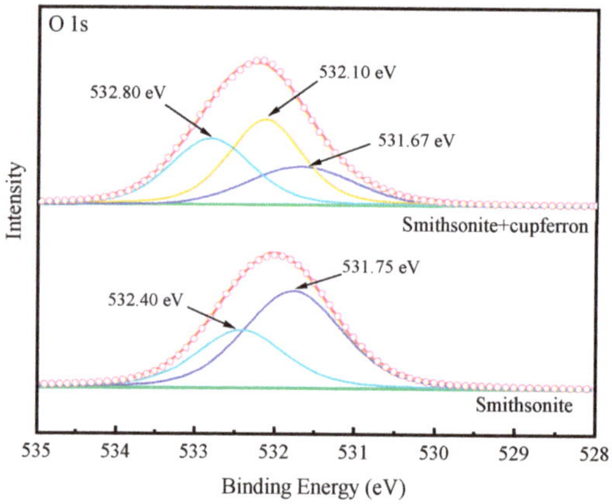

Figure 14. O 1s XPS analysis of cupferron-treated smithsonite.

Table 3 illustrates the relative atomic concentrations on the surface of smithsonite and calcite before and after cupferron treatment. After the surface of smithsonite was treated with cupferron, the relative abundance of C and N atoms increased compared with that of bare smithsonite, while the relative content of O and Zn atoms decreased. Compared with the surface of calcite, the overall relative atomic content changed little—less than 0.1%. This indicated that cupferron was easily adsorbed on the surface of smithsonite and interacted with the surface Zn sites, resulting in a decrease in the relative content of Zn. The surface of calcite hardly adsorbed cupferron. Figure 15 shows the high-resolution spectra of Ca 2p. The binding energies of 347.20 eV and 350.70 eV are ascribed to Ca 2p1/2 and Ca 2p3/2. When calcite was treated with cupferron, the two peaks of Ca 2p3/2 and Ca 2p1/2 merely decreased, and the peak only shifted to 0.05 eV. From this, we concluded that cupferron chelated with smithsonite does not interact with calcite.

Table 3. The relative atomic content changes on the surface of smithsonite and calcite.

Samples	Atomic Concentration, %				
	C	O	Zn	Ca	N
Smithsonite	30.74	48.19	21.07	—	—
Smithsonite + Cupferron	30.87	47.52	20.83	—	0.78
Calcite	31.32	49.27	—	19.41	—
Calcite + Cupferron	31.38	49.21	—	19.36	0.05

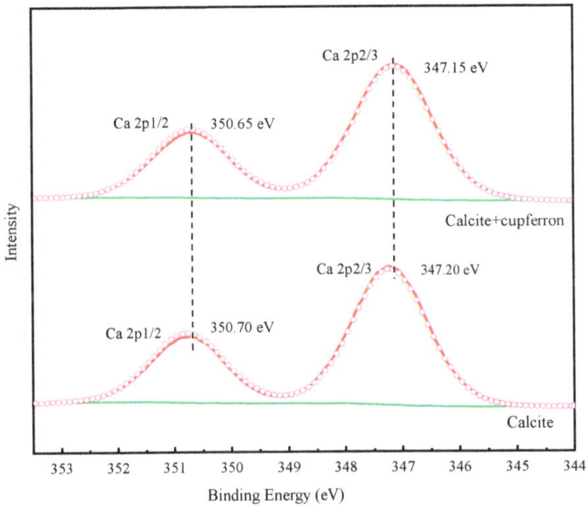

Figure 15. Ca 2p XPS analysis of calcite and cupferron-treated calcite.

3.6. The Mechanism of the Flotation of Smithsonite with Cupferron as the Collector

From these flotation experiments and characterization results, the adsorption model of cupferron adsorption on the smithsonite surface can be constructed. As shown in Figure 16, compared with calcite, the cupferron was more easily adsorbed on the surface of smithsonite and altered the surface hydrophobicity of smithsonite. Infrared results showed that new chemical bond vibration peaks were observed in the spectrum of cupferron-treated smithsonite. These vibration peaks belonged to the characteristic peaks of cupferron, including N-O functional group and C-H functional group of the benzene ring. The Zn-O bond on the surface of smithsonite also shifted. A new peak N 1s was found in the XPS analysis results. The infrared results were further validated by the high-resolution spectra of Zn 2p and O 1s. The O 1s spectra of the cupferron-treated smithsonite samples were fitted by three well-separated peaks, the O peaks in zinc carbonate and zinc hydroxide were shifted, and there was an extra peak that could be ascribed to the N-O group. The results clearly indicated that cupferron as a chelating reagent can form a five-membered ring complex with metal ions [26–28]. On the basis of these findings, cupferron was mostly chelated with the smithsonite surface zinc sites to form the insoluble species.

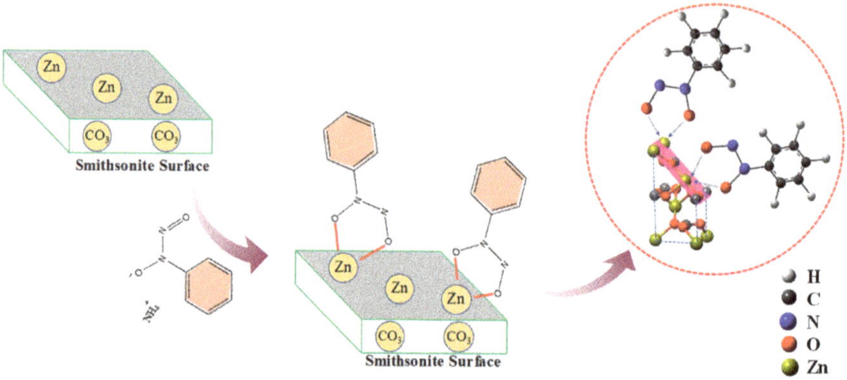

Figure 16. The adsorption mechanism of cupferron on the surface of smithsonite and calcite.

4. Conclusions

This study demonstrated the effective separation of smithsonite from calcite using cupferron as a chelating collector. Mineral flotation and adsorption experiments revealed significant differences in the flotation behavior of calcite and smithsonite in the presence of cupferron. Smithsonite exhibited a higher adsorption capacity for cupferron, resulting in a recovery of 79.08% at a pH of 8.0. Zeta potential measurements indicated a greater decrease in zeta potential for smithsonite compared to calcite when treated with cupferron, suggesting a stronger effect of cupferron on smithsonite. Furthermore, IR and XPS spectra analysis demonstrated that cupferron chemically adsorbed onto the surface of smithsonite, primarily interacting with Zn^{2+} ions to form hydrophobic chelate complexes. These findings highlighted cupferron as a potential collector for smithsonite flotation, offering an alternative to sulfidation flotation methods and improving selectivity while mitigating environmental pollution.

Author Contributions: Formal analysis, investigation, writing–original draft, Q.W.; formal analysis, data curation, writing—review and editing, Y.C. and X.W.; investigation, data curation, formal analysis, L.S.; methodology, project administration, resources, supervision, writing—review and editing, Y.Q. and M.X.; conceptualization, methodology, funding acquisition, G.L. and W.S. All authors have read and agreed to the published version of the manuscript.

Funding: All authors of this work acknowledged the financial support from the National Key Technology Research and Development Program of China (2020YFC1909202), and the Major Science and Technology Program of Yunnan Province, China (202202AB080012).

Data Availability Statement: The data supporting the findings of this study are available from the corresponding author upon reasonable request.

Acknowledgments: We sincerely thank the editor and reviewers for their constructive comments which helped in improving our paper.

Conflicts of Interest: The authors declare no conflict of interest.

References

1. Tao, M.; Zhang, X.; Wang, S.; Cao, W.; Jiang, Y. Life Cycle Assessment on Lead–Zinc Ore Mining and Beneficiation in China. *J. Clean. Prod.* **2019**, *237*, 117833. [CrossRef]
2. Abkhoshk, E.; Jorjani, E.; Al-Harahsheh, M.S.; Rashchi, F.; Naazeri, M. Review of the Hydrometallurgical Processing of Non-Sulfide Zinc Ores. *Hydrometallurgy* **2014**, *149*, 153–167. [CrossRef]
3. Balarini, J.C.; Polli, L.d.O.; Miranda, T.L.S.; de Castro, R.M.Z.; Salum, A. Importance of Roasted Sulphide Concentrates Characterization in the Hydrometallurgical Extraction of Zinc. *Miner. Eng.* **2008**, *21*, 100–110. [CrossRef]
4. Ejtemaei, M.; Gharabaghi, M.; Irannajad, M. A Review of Zinc Oxide Mineral Beneficiation Using Flotation Method. *Adv. Colloid Interface Sci.* **2014**, *206*, 68–78. [CrossRef] [PubMed]

5. Zhao, L.; Liu, W.; Duan, H.; Wang, X.; Fang, P.; Liu, W.; Zhou, X.; Shen, Y. Design and Selection of Flotation Collectors for Zinc Oxide Minerals Based on Bond Valence Model. *Miner. Eng.* **2021**, *160*, 106681. [CrossRef]
6. Chen, Y.; Liu, M.; Chen, J.; Li, Y.; Zhao, C.; Mu, X. A Density Functional Based Tight Binding (DFTB+) Study on the Sulfidization-Amine Flotation Mechanism of Smithsonite. *Appl. Surf. Sci.* **2018**, *458*, 454–463. [CrossRef]
7. Hosseini, S.H.; Forssberg, E. Physicochemical Studies of Smithsonite Flotation Using Mixed Anionic/Cationic Collector. *Miner. Eng.* **2007**, *20*, 621–624. [CrossRef]
8. Irannajad, M.; Ejtemaei, M.; Gharabaghi, M. The Effect of Reagents on Selective Flotation of Smithsonite–Calcite–Quartz. *Miner. Eng.* **2009**, *22*, 766–771. [CrossRef]
9. Mehdilo, A.; Irannajad, M.; Zarei, H. Smithsonite Flotation from Zinc Oxide Ore Using Alkyl Amine Acetate Collectors. *Sep. Sci. Technol.* **2014**, *49*, 445–457. [CrossRef]
10. Chen, Y.; Zhang, G.; Wang, M.; Shi, Q.; Liu, D.; Li, Q. Utilization of Sodium Carbonate to Eliminate the Adverse Effect of Ca^{2+} on Smithsonite Sulphidisation Flotation. *Miner. Eng.* **2019**, *132*, 121–125. [CrossRef]
11. Feng, Q.; Wen, S.; Bai, X.; Chang, W.; Cui, C.; Zhao, W. Surface Modification of Smithsonite with Ammonia to Enhance the Formation of Sulfidization Products and Its Response to Flotation. *Miner. Eng.* **2019**, *137*, 1–9. [CrossRef]
12. Lan, Z.; Lai, Z.; Zheng, Y.; Lv, J.; Pang, J.; Ning, J. Thermochemical Modification for the Surface of Smithsonite with Sulfur and Its Flotation Response. *Miner. Eng.* **2020**, *150*, 106271. [CrossRef]
13. Li, C.; Bai, S.; Ding, Z.; Yu, P.; Wen, S. Visual MINTEQ Model, ToF–SIMS, and XPS Study of Smithsonite Surface Sulfidation Behavior: Zinc Sulfide Precipitation Adsorption. *J. Taiwan Inst. Chem. Eng.* **2019**, *96*, 53–62. [CrossRef]
14. Luo, B.; Liu, Q.; Deng, J.; Yu, L.; Lai, H.; Song, C.; Li, S. Characterization of Sulfide Film on Smithsonite Surface during Sulfidation Processing and Its Response to Flotation Performance. *Powder Technol.* **2019**, *351*, 144–152. [CrossRef]
15. Cai, J.; Su, C.; Ma, Y.; Yu, X.; Peng, R.; Li, J.; Zhang, X.; Fang, J.; Shen, P.; Liu, D. Role of Ammonium Sulfate in Sulfurization Flotation of Azurite: Inhibiting the Formation of Copper Sulfide Colloid and Its Mechanism. *Int. J. Min. Sci. Technol.* **2022**, *32*, 575–584. [CrossRef]
16. Liu, M.; Chen, J.; Chen, Y.; Zhu, Y. Interaction between Smithsonite and Carboxyl Collectors with Different Molecular Structure in the Presence of Water: A Theoretical and Experimental Study. *Appl. Surf. Sci.* **2020**, *510*, 145410. [CrossRef]
17. Liu, W.; Wang, Z.; Wang, X.; Miller, J.D. Smithsonite Flotation with Lauryl Phosphate. *Miner. Eng.* **2020**, *147*, 106155. [CrossRef]
18. Wang, Z.; Xu, L.; Wang, J.; Wang, L.; Xiao, J. A Comparison Study of Adsorption of Benzohydroxamic Acid and Amyl Xanthate on Smithsonite with Dodecylamine as Co-Collector. *Appl. Surf. Sci.* **2017**, *426*, 1141–1147. [CrossRef]
19. Zhao, L.; Liu, W.; Duan, H.; Yang, T.; Li, Z.; Zhou, S. Sodium Carbonate Effects on the Flotation Separation of Smithsonite from Quartz Using N,N′-Dilauroyl Ethylenediamine Dipropionate as a Collector. *Miner. Eng.* **2018**, *126*, 1–8. [CrossRef]
20. Dong, L.; Qiao, L.; Zheng, Q.; Shen, P.; Qin, W.; Jiao, F.; Liu, D. Enhanced Adsorption of Citric Acid at the Calcite Surface by Adding Copper Ions: Flotation Separation of Scheelite from Calcite. *Colloids Surf. A Physicochem. Eng. Asp.* **2023**, *663*, 131036. [CrossRef]
21. Liu, C.; Zhang, W.; Song, S.; Li, H.; Liu, Y. Flotation Separation of Smithsonite from Calcite Using 2-Phosphonobutane-1,2,4-Tricarboxylic Acid as a Depressant. *Powder Technol.* **2019**, *352*, 11–15. [CrossRef]
22. Luo, Y.; Zhang, G.; Mai, Q.; Liu, H.; Li, C.; Feng, H. Flotation Separation of Smithsonite from Calcite Using Depressant Sodium Alginate and Mixed Cationic/Anionic Collectors. *Colloids Surf. A Physicochem. Eng. Asp.* **2020**, *586*, 124227. [CrossRef]
23. Luo, Y.; Zhang, G.; Li, C.; Mai, Q.; Liu, H.; Zhou, H.; Shi, Q. Flotation Separation of Smithsonite from Calcite Using a New Depressant Fenugreek Gum. *Colloids Surf. A Physicochem. Eng. Asp.* **2019**, *582*, 123794. [CrossRef]
24. Marabini, A.M.; Ciriachi, M.; Plescia, P.; Barbaro, M. Chelating Reagents for Flotation. *Miner. Eng.* **2007**, *20*, 1014–1025. [CrossRef]
25. Muthuswami, S.V.; Vijayan, S.; Woods, D.R. Flotation of Uranium from Uranium Ores in Canada: Part II—Cupferron Adsorption on Uranium Oxide, Quartz, Illite and a Uranium Ore from Elliot Lake. *Can. J. Chem. Eng.* **1985**, *63*, 650–661. [CrossRef]
26. Prabhakar, S.; Khangaonkar, P.R. Flotation and Adsorption Studies of Chalcopyrite with Cupferron. *Int. J. Miner. Process.* **1982**, *9*, 87–95. [CrossRef]
27. Sheng, Q.; Yang, B.; Cao, S.; Yin, W.; Sun, H.; Ma, Y.; Chen, K. Adsorption of Cupferron on Malachite (−201) Surface and Implication for Flotation. *Miner. Eng.* **2021**, *169*, 106954. [CrossRef]
28. Bahri, Z.; Rezai, B.; Kowsari, E. Evaluation of Cupferron on the Selective Separation of Gallium from Aluminum by Flotation: The Separation Mechanism. *Miner. Eng.* **2016**, *98*, 194–203. [CrossRef]
29. Bottei, R.S.; Schneggenburger, R.G. Thermal and Spectral Study of Some Divalent Metal Chelates of Cupferron and Dicupferron. *J. Inorg. Nucl. Chem.* **1970**, *32*, 1525–1545. [CrossRef]
30. Shi, Q.; Zhang, G.; Feng, Q.; Deng, H. Effect of Solution Chemistry on the Flotation System of Smithsonite and Calcite. *Int. J. Miner. Process.* **2013**, *119*, 34–39. [CrossRef]
31. Shi, Q.; Zhang, G.; Feng, Q.; Ou, L.; Lu, Y. Effect of the Lattice Ions on the Calcite Flotation in Presence of Zn(II). *Miner. Eng.* **2013**, *40*, 24–29. [CrossRef]
32. Thakur, N.V.; Kartha, V.B.; Kanekar, C.R.; Marathe, V.R. Infrared Spectra of Cupferron and Some Rare Earth Cupferrates. *J. Inorg. Nucl. Chem.* **1972**, *34*, 2831–2836. [CrossRef]
33. Frost, R.L.; Reddy, B.J.; Wain, D.L.; Hales, M.C. An Application of near Infrared Spectroscopy to the Study of Carbonate Minerals—Smithsonite, Rhodochrosite, Sphaerocobaltite and Cadmium Smithsonite. *J. Near Infrared Spectrosc.* **2006**, *14*, 317–324. [CrossRef]

34. Hales, M.C.; Frost, R.L. Synthesis and Vibrational Spectroscopic Characterisation of Synthetic Hydrozincite and Smithsonite. *Polyhedron* **2007**, *26*, 4955–4962. [CrossRef]
35. Ebrahimiasl, S.; Zakaria, A.; Kassim, A.; Norleha Basri, S. Novel Conductive Polypyrrole/Zinc Oxide/Chitosan Bionanocomposite: Synthesis, Characterization, Antioxidant, and Antibacterial Activities. *IJN* **2014**, *10*, 217. [CrossRef]
36. Wahab, R.; Ansari, S.G.; Kim, Y.S.; Song, M.; Shin, H.-S. The Role of PH Variation on the Growth of Zinc Oxide Nanostructures. *Appl. Surf. Sci.* **2009**, *255*, 4891–4896. [CrossRef]
37. Chen, Y.; Guo, X.; Chen, Y. Using Phytic Acid as a Depressant for the Selective Flotation Separation of Smithsonite from Calcite. *Sep. Purif. Technol.* **2022**, *302*, 122104. [CrossRef]
38. Xie, X.; Li, B.; Xie, R.; Tong, X.; Li, Y.; Zhang, S.; Li, J.; Song, Q. Al^{3+} Enhanced the Depressant of Guar Gum on the Flotation Separation of Smithsonite from Calcite. *J. Mol. Liq.* **2022**, *368*, 120759. [CrossRef]
39. Liu, Z.; Teng, F. Understanding the Correlation of Crystal Atoms with Photochemistry Property: $Zn_5(OH)_6(CO_3)_2$ vs. $ZnCO_3$. *ChemistrySelect* **2018**, *3*, 8886–8894. [CrossRef]
40. Zhu, Y.; Yu, X.; Pan, S.; Nong, P.; Kong, Q.; Wang, X.; Zhang, L.; Tan, S.; Zhu, Z. Dissolution of the Smithsonite–Rhodochrosite ($ZnCO_3$-$MnCO_3$) Solid Solutions in Aqueous Solution at 25 °C. *Chem. Geol.* **2022**, *602*, 120886. [CrossRef]
41. Gomez-Bolivar, J.; Mikheenko, I.P.; Orozco, R.L.; Sharma, S.; Banerjee, D.; Walker, M.; Hand, R.A.; Merroun, M.L.; Macaskie, L.E. Synthesis of Pd/Ru Bimetallic Nanoparticles by *Escherichia coli* and Potential as a Catalyst for Upgrading 5-Hydroxymethyl Furfural Into Liquid Fuel Precursors. *Front. Microbiol.* **2019**, *10*, 1276. [CrossRef] [PubMed]

minerals MDPI

Article

Process Mineralogy Characteristics and Flotation Optimization of a Low-Grade Oxidized Lead and Zinc Ore from Lanping Mine

Anmei Yu [1,2], Zhan Ding [1,2], Jiaqiao Yuan [1,2], Qicheng Feng [1,2,3], Shuming Wen [1,2] and Shaojun Bai [1,2,3,*]

1 Faculty of Land Resource Engineering, Kunming University of Science and Technology, Kunming 650093, China; yuanmei199805@126.com (A.Y.); kustdingzhan@126.com (Z.D.); yuankust@126.com (J.Y.); fqckmust@126.com (Q.F.); shmwen@126.com (S.W.)

2 State Key Laboratory of Complex Nonferrous Metal Resources Clean Utilization, Kunming University of Science and Technology, Kunming 650093, China

3 Yunnan Key Laboratory of Green Separation and Enrichment of Strategic Mineral Resources, Kunming 650093, China

* Correspondence: 20110055@kust.edu.cn; Tel.: +86-138-8840-0155

Abstract: The beneficiation of low-grade oxidized lead and zinc ore from the Lanping mine has attracted extensive interest in the mineral processing field due to the extremely rich resource reserves and the challenge in lead and zinc recovery. This study's objective is to analyze the process mineralogy characteristics and to conduct the flotation optimization of this ore. Mineralogy parameters of the ore, such as mineral composition, mineral liberation degree, and intergrowth relationship, were investigated using an X-ray fluorescence spectrometer (XRF), an X-ray powder diffractometer (XRD), and an electronic probe microanalyzer (EPMA). The effect of the main experimental factors on the lead and zinc flotation is determined and compared. The results confirm that the low content of lead and zinc and the complex co-occurrence relationships between minerals bring a challenge to the efficient separation of the ore. Furthermore, a flow sheet consisting of "lead preferred flotation-zinc flotation" is performed, and an all-open flotation process consisted of "two-times lead rougher-one-time lead cleaner-two-times zinc rougher-one-time zinc cleaner" is finally employed for the beneficiation of this ore with the aids of mixed depressants and cationic-anionic collectors for zinc recovery. The Pb grade in the lead rougher concentrate is 2.83%, and the Pb recovery is 57.56%. The Zn grade reaches 28.64% with a recovery of 83.45%. Thus, the findings provide important technical supports for the processing of similar ores.

Keywords: low-grade oxidized lead and zinc ore; process mineralogy; selective flotation; all-open flotation flowsheet; cationic-anionic collectors

1. Introduction

Zinc and lead enjoy tremendous importance owing to their wide application in industries. In the past few decades, more than eighty percent of zinc and lead originated from sulfide minerals (e.g., sphalerite, galena) because of their good hydrophobicity. With the fast consumption of zinc lead sulfide ores, the beneficiation of oxidized lead and zinc ores has assumed important significance recently because there are numerous unexploited or abandoned reserves of these ores in the world [1]. It is estimated that zinc-lead ores form 0.13% of the Earth's crust [2]. Lead and zinc mineral resources are mainly distributed in the USA, Canada, China, Australia, Mexico, Germany, and Poland [3]. For instance, Yunnan Jinding Zinc Industry Co., LTD. (located in Lanping, China) has the world's fifth- largest lead and zinc mineral resources. Specifically, the unexploited lead and zinc resources reserve reaches 39.88 million tons, and the potential economic value is up to 50 billion yuan. It has become the primary problem that restricts the sustainable development of the Jinding zinc industry.

Process mineralogy combines quantitative (and qualitative) mineralogical techniques with flotation test work [4]. Modern process mineralogy has been devising significant advances in methodology and data construction since the mid-1980s [5]. Several studies have remarked on the significance of process mineralogy in mineral processing [6]. It is considered an essential requirement for characterizing the surface morphology and grain size distribution scale of ores [7]. Through process mineralogy research on samples, the rationality of the production process and reagent conditions can be diagnosed, and a basis for optimizing the production process [8]. Thus, it is vital to identify the process mineralogy characteristics of low-grade oxidized lead and zinc ore from the Lanping mine, which will provide an important basis for the formulation of the subsequent beneficiation process.

Roasting, leaching, microbial hydrometallurgy, and hydrothermal vulcanization could obtain desirable technical indexes to recover zinc and lead from low-grade zinc oxide deposits [9–12]; however, the application of these process routes has been usually uneconomical. Froth flotation methods are essential to the recovery of zinc and lead from these low-grade zinc oxide ores [13]. Of course, flotation reagents and flotation flow-sheets were the keys to addressing this obstacle originating from low-grade, highly oxidized properties and the complex co-occurrence relationships between minerals. Traditionally, the flotation of zinc and lead ores with fatty amines was used to enrich the target minerals. This approach required large amounts of reagents and had poor selectivity when the raw ore contained carbonate minerals. More importantly, the fatty amines usually fail to recover the oxidized zinc mineral effectively if a multitude of slimes are present in the pulp because the slimes reduce the collector selectivity and lead to difficulty in defoaming. Additionally, sulfidation-xanthate flotation has obvious merits for the treatment of lead zinc oxidized ores that contain plenty of slimes. However, the flotation recovery of zinc was still not acceptable due to the poor collecting capacity of xanthates [14–16]. With these in mind, the combination of mixed depressants and cationic-anionic collectors was proposed as an efficient reagent regime to beneficiate zinc and lead from the ore. Moreover, an all-open flotation process was employed to avoid the undesirable effects of slimes.

The aim of this paper is to develop an efficient flotation process to selectively separate zinc and lead from high-oxidized lead zinc ore. Firstly, the process mineralogy of the ore was studied via an X-ray fluorescence spectrometer (XRF), an X-ray powder diffractometer (XRD), and an electronic probe microanalyzer (EPMA) to provide a theoretical basis for a suitable flotation process. Afterwards, the systematic flotation experiments were performed to obtain the optimum flotation separation conditions. The present study provided important technical support for the processing of similar ores.

2. Materials and Methods

2.1. Ore Sample and Reagents

The raw material was obtained from an open storage yard in Lanping, Yunnan, China. It was crushed to 1~2 mm and mixed to achieve a uniform composition. A part of the representative sample was ground to below 75 μm particle size using a three-end grinder (model XPM-Φ120 × 8, Jiangxi Weiming Machinery Equipment Co., Ltd., Jiangxi, China) for X-ray fluorescence spectrometer (XRF), and X-ray powder diffractometer (XRD) detection. Some samples were solidified by an epoxy resin and a solidifying agent, and were then polished and sprayed with gold to guarantee the conductivity of the sample surface for the electron probe X-ray microanalysis (EPMA). In addition, the other part of the crushed product was ground by a rod mill (Wuhan Prospecting Machinery Factory, Wuhan, China). The grinding time was set to 40 s, 60 s, 80 s, 100 s, and 120 s, and the products were wet sieved by a 74 μm sieve, and the grinding fineness was determined. Figure 1 shows the relationship between grinding fineness and grinding time.

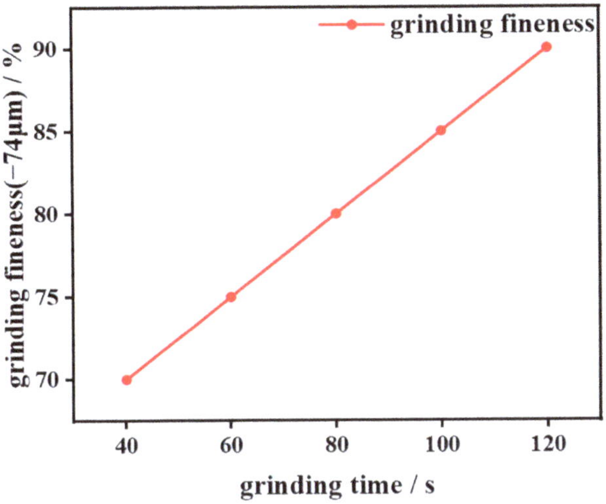

Figure 1. The relationship between grinding fineness and grinding time.

The sulphidizing agent (Na_2S), the pH adjuster and dispersant (Na_2CO_3), and the frother (pine oil) were obtained from the Tianjin Hedong Hongyan Reagent Factory, Tianjin, China. The depressants (DCCH is a kind of organic dispersant that is rich in multiple hydroxyl and carboxyl functional groups; sodium silicate) were purchased from Tianjin Fengboat Chemical Reagent Technology Co., Ltd., Tianjin, China. The collectors (isoamyl xanthate and cations and anions collectors) were purchased from Fuchen Chemical Reagent Co., Ltd., Tianjin, China. All the chemical reagents used in this study were of analytical grade, and tap water was used in all experiments.

2.2. Characterization Methods

An X-ray fluorescence spectrometer (XRF, EDX 6600, suzhou three value precision instrument Co., LTD, Jiangsu, China) and X-ray diffraction (XRD, ARL EQUNINX 3000 X) were used for chemical elemental quantitative analysis and mineral phase identification, respectively. XRF with a rhodium-anode tube was used for the quantitative analysis of the chemical elements at 4 kW. The main operating parameters of XRD were as follows: Cu target radiation, a scanning range of 0°–80°, and a scanning speed of 12°/min. The microstructures, morphologies, and elemental composition of the samples were evaluated by optical microscope (ZZJ-2030, Shenzhen Zhongzheng Instrument Co., LTD, Guangzhou, Guangdong) and electron probe X-ray microanalyzer (EPMA, JXA-iSP100, Tokyo, Japan), which was operated with a 15 kV accelerating voltage, a 20 nA beam current, and a 5 μm beam diameter. The contents of lead and zinc were analyzed through the a chemical titration precipitation method.

2.3. Flotation Tests

Flotation experiments were conducted in a flotation cell (model XFD III, Jiangxi Weiming Machinery Equipment Co., Ltd., Jiangxi, China). To create a pulp density of ~28%, 400 g of lead-zinc oxide ore and a specific dosage of water were added to the flotation cell with a capacity of 1.5 L. The flotation machine speed was set at 1900 r/min. After that, the dispersant (Na_2CO_3), the depressants (DCCH and sodium silicate), the sulphidizing agent (Na_2S), the collectors (isoamyl xanthate, cations, and anions collectors), and a frother (pine oil) were successively added into the pulp. Each reagent was conditioned for 2–5 min. The flotation time was fixed for 8 min at room temperature. Flotation concentrates and tailings were dried, weighed, sampled, and tested for grades of target elements. Recovery of the concentrate was calculated and treated as a response to evaluate the flotation index. In addition, the fluctuation ranges of recovery rate and grade in each factor of the flotation condition experiment were contrastively analyzed to explore the influence degree of test factors on the flotation indexes. They were calculated by the following formulas.

$$\varepsilon = \varepsilon_{max} - \varepsilon_{min} \tag{1}$$

$$\beta = \beta_{max} - \beta_{min} \tag{2}$$

Here, ε_{max} represented the maximum recovery rate of each factor, and ε_{min} represented the minimum recovery rate for each factor. β_{max} was the maximum grade rate of each factor, and β_{min} was the minimum grade rate for each factor.

3. Results and Discussion

3.1. Process Mineralogy Analysis

3.1.1. Mineral Composition and Element Occurrence

Data for the chemical compositions and chemical phases of the ore are provided in Tables 1 and 2, respectively. From Table 1, the content of lead and zinc in the samples was relatively low [17]. The Zn content in the sample was 7.4%, and that of Pb was 0.84% by chemical composition analysis. Additionally, the contents of SiO_2 and CaO in the raw ore were 18.96% and 26.02%, respectively, indicating that the ore contained silicate and carbonate minerals. It is worth noting that the content of the harmful element As was only 0.042%, which could be negligible. Thus, Zn and Pb were the main recoverable elements from the oxide ore. From Table 2, a total content of 89.05% of zinc minerals existed as oxide; the content of lead oxide minerals was 51.19%, which was in the form of cerussite. In addition, 8.92% of the zinc minerals existed as sulfides. The distribution rate of zinc in other forms was only 2.03%. The above results indicated that the ore sample was a low-grade lead-zinc oxidized ore with high contents of iron, silicon, and calcium.

Table 1. Chemical composition of lead-zinc oxide ore (mass fraction, %).

Element	Pb	Zn	K_2O	Au *	Cu
Content	0.84	7.4	0.81	0.06	0.0099
Element	Fe	As	CaO	MgO	SiO_2
Content	5.37	0.042	26.02	0.79	18.96

* Unit g/t.

Table 2. The chemical phases of ore.

Element	Elemental Phase	Content (%)	Distribution Ratio (%)
Pb	Cerussite	0.43	51.19
	Galena	0.12	14.29
	Anglesite	0.16	19.05
	Plumbojarosite and others	0.13	15.48
	Total	0.84	100.00
Zn	Sulfides	0.66	8.92
	Zinc sulfate	0.05	0.68
	Oxide	6.59	89.05
	Franklinite and others	0.10	1.35
	Total	7.40	100.00

XRD results (as depicted in Figure 2) indicated that smithsonite and cerussite were the main valuable metallic minerals, and calcite ($CaCO_3$) and quartz (SiO_2) were the main gangue minerals. The high oxidation rates with multi-metals and large quantities of calcite and quartz minerals would increase the difficulty of ore beneficiation [18,19]. Thus, XRD results were consistent with the chemical composition results, and the froth flotation would provide easy access to the recovery of zinc and lead minerals.

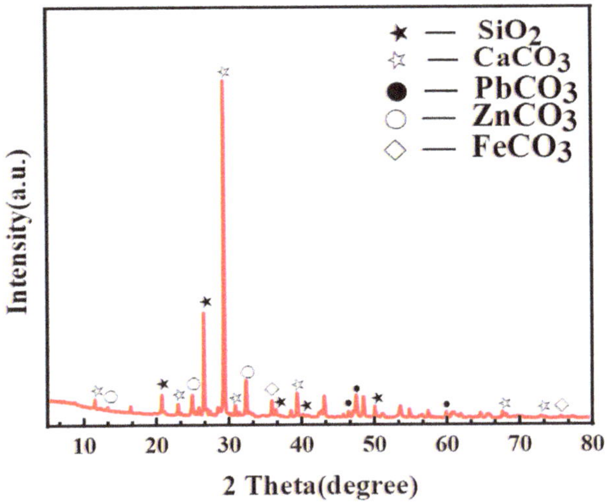

Figure 2. XRD results for lead-zinc oxide ore.

The XRD quantitative test was employed to analyze mineral composition, which is shown in Table 3. Mineral composition of lead-zinc oxide ore The results showed that the main lead minerals were cerussite and galena, while the main zinc mineral was smithsonite and sphalerite. Gangue minerals contained silicate minerals (quartz, muscovite, kaolinite, potassium feldspar, etc.), carbonate minerals (calcite, dolomite, etc.), and metal minerals (pyrite, siderite, and goethite). Among them, the content of siderite was as high as 9.85%. A large quantity of gangue minerals with low hardness was negative for flotation. Consequently, the control of grind fineness was inevitable to prevent over-grinding and to create a desirable floatation environment.

Table 3. Mineral composition of lead-zinc oxide ore.

Mineral	Molecular Formula	Content/%	Mineral	Molecular Formula	Content/%
Quartz	SiO_2	15.02	Smithsonite	$ZnCO_3$	10.49
Calcite	$CaCO_3$	40.97	Siderite	$FeCO_3$	9.85
Dolomite	$CaMg(CO_3)_2$	3.07	Cerussite	$PbCO_3$	0.74
Sphalerite	ZnS	0.88	Pyrite	FeS_2	1.09
Galena	PbS	0.44	Hemimorphite	$Zn_4Si_2O_7(OH)_2(H_2O)$	2.42
Goethite	FeO(OH)	1.27	Barite	$BaSO_4$	1.36
Celestine	$SrSO_4$	1.46	Gypsum	$CaSO_4{\cdot}2H_2O$	2.08
Muscovite	$KAl_2Si_3AlO_{10}(OH)_2$	0.82	Kaolinite	$Al_2Si_2O_5(OH)_4$	0.96
Potassium feldspar	$KAlSi_3O_8$	2.08	Others	-	5.00

The ore was further analyzed by EPMA, and the results are shown in Figure 3. There were a lot of carbon and oxygen elements in the ore, and their distributions were consistent. Sulfur was almost absent and was consistent with the aforementioned results. In addition, the distribution of iron and zinc elements exhibited overlap, indicating that siderite and smithsonite might be co-associated in this ore. The distribution of silicon and calcium was similar, indicating some of the quartz and calcite might be associated together. The amounts of lead, copper, and magnesium were small, but the distribution was uniform. Thus, low content of lead and zinc and complex co-occurrence relationships between minerals, especially, iron and zinc, would bring a challenge to the efficient separation of the ore.

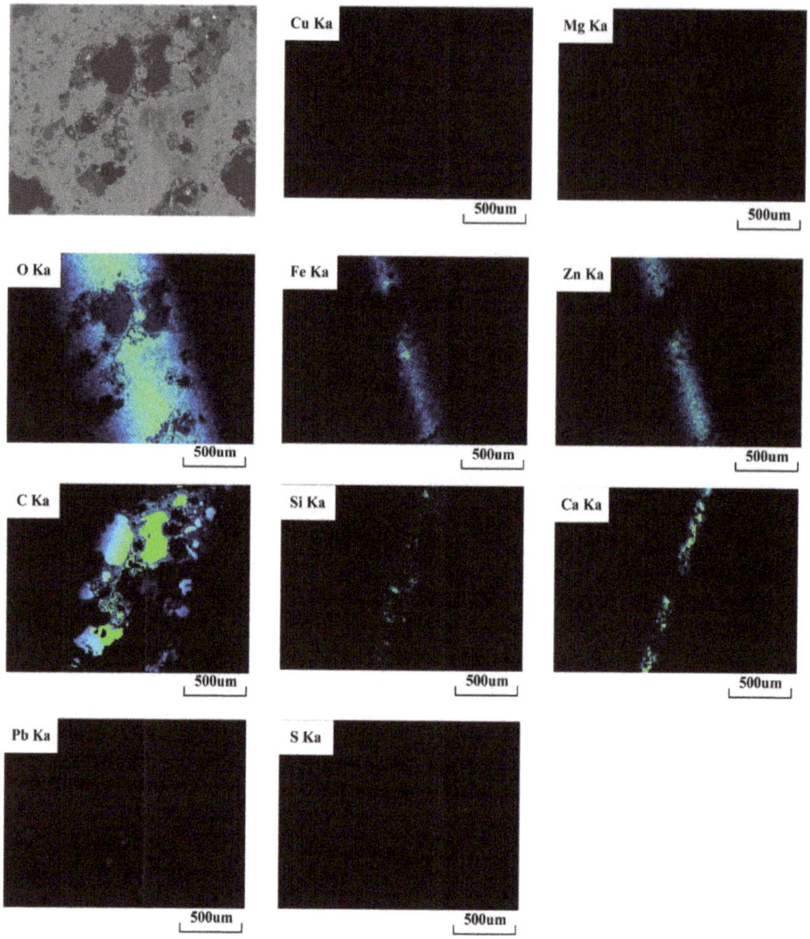

Figure 3. EPMA elements mapping of Pb-Zn oxide.

3.1.2. Zinc Minerals Embedded Characteristic

Zn was observed in the form of smithsonite and sphalerite. Smithsonite observed under the optical microscope was gray. Because of the dissemination with ferric hydroxide, most particulates of the smithsonite particles in the ore mainly had a microgranular particle size, while a small percentage of that had the false appearance of sphalerite (Figure 4a). Its single crystal average was within 0.1 mm, showing a porous and colloidal structure. Sphalerite was also gray with a strong metallic luster under the microscope. Some sphalerite occurring as stripes and irregular ellipses formed with the size of 0.002–0.3 mm (Figure 4b). Most of the smithsonite and sphalerite had a close intergrowth relation with sphalerite, pyrite, and gangue minerals, embedding particle size from 0.05 to 0.25 mm. From Figure 4c, part of the sphalerite was crumpled, resulting in many fissures or cracks. They were filled and metasomatized by gangue and pyrite. Part of the pyrite is immersed along non-metallic mineral fissures, forming vein-like structures. A portion of the sphalerite was wrapped by marcasite and gangue with a size of 0.1–0.15 mm. In addition, part of the sphalerite with smithsonite intergrowth existed in the form of pellet aggregate, which was generally 0.2–0.3 mm in diameter. Some of the smithsonite and sphalerite were fine-grained irregular shape aggregates distribution. The sphalerite had multiple blocky structures and was band shaped (Figure 4d). Given these results, it was essential to realize the particulate's liberation of zinc-bearing minerals and to avoid the over-grinding of gangue minerals for the recovery of zinc.

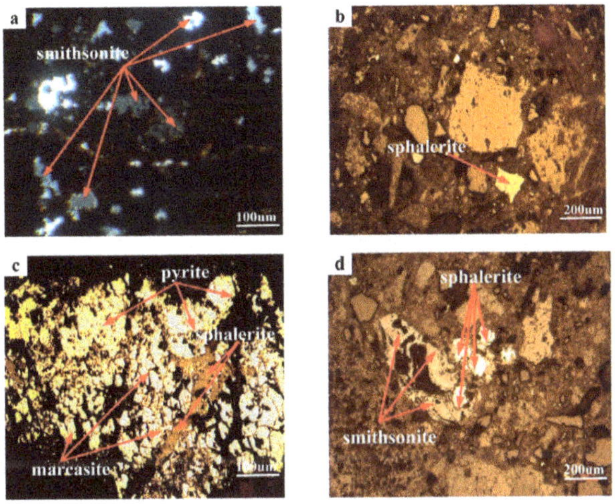

Figure 4. The particulate and distribution form of smithsonite and sphalerite. ((**a**): smithsonite particulate; (**b**): sphalerite particulate; (**c**): distribution form of sphalerite, marcasite, and pyrite; (**d**): distribution form of smithsonite and sphalerite. Viewed under an optical microscope).

3.1.3. Lead Minerals Embedded Characteristic

Pb were observed in the form of galena and cerussite. Galena was opaque black under the optical microscope. Part of the galena existed as particulates and most of them were triangular in cross-section and had a distributed particle size from 0.20 to 1.50 mm (Figure 5a). Its grain size was much larger than cerussite with a poorer liberation. The color of the cerussite was pale yellow and it had almost no particulates. The existence was usually secondary metasomatism along the edge of the galena grains, with a rhythmic crust structure (Figure 5b,c). These aggregates had an embedded particle size from 0.10 to 1.60 mm. Furthermore, galena was mainly compactly associated with sphalerite and pyrite. This was already demonstrated in Figure 5d. The embedding particle size of galena in the ore was mainly between 0.20 and 1.00 mm. The content of cerussite and galena in this ore was extremely low and was difficult to recover.

3.1.4. Gangue Minerals Embedded Characteristic

Calcite was the main gangue in the ore. It presented a colorless fine microcrystalline granular structure under the optical microscope. Most of the grains were more than 1 mm, forming an embedded structure with each other, as shown in Figure 6a. Quartz was another major gangue mineral in the ore. It was observed to be colorless, with allomorphic granular or semi-euhedral columns. The particle size was mainly between 0.2 and 0.6 mm. Some of the quartz existed in the form of a particulate in the structure of drusy breccia (Figure 6b) with particles >2 mm. This part of smithsonite was difficult to liberate absolutely and to separate, leading to the Si enriching in the flotation concentrate. In addition, some micrite was crumpled, resulting in numerous fissures or cracks. These fissures and cracks were subsequently filled and metasomatized by quartz and feldspar (Figure 6c). A portion of the smithsonite was filled with cement among the quartz grains (Figure 6d). Thus, the gangue-embedded characteristic further indicated multitude of slimes would be formed and affect the concentrate quality.

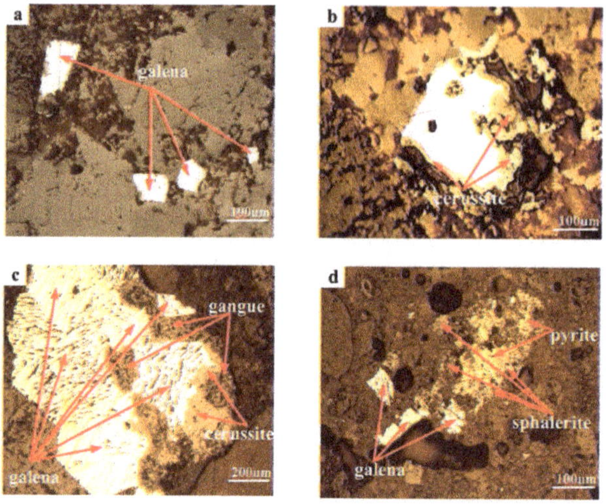

Figure 5. The particulate and distribution form of galena and cerussite. ((**a**): galena particulate; (**b**): cerussite particulate; (**c**): distribution form of gangue, galena, and cerussite; (**d**): distribution form of pyrite, galena and sphalerite. Viewed under an optical microscope).

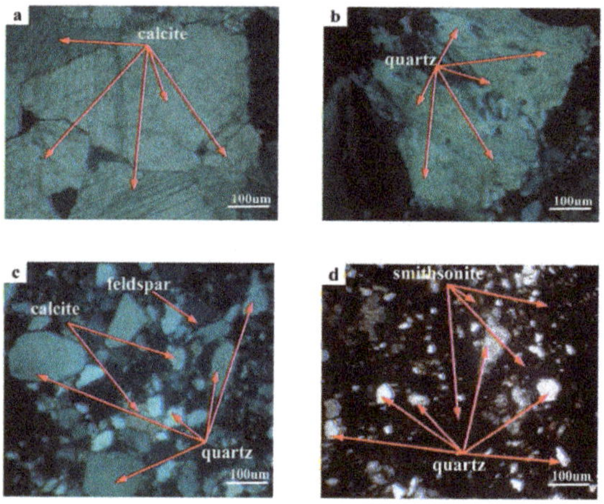

Figure 6. The particulate and distribution form of calcite and quartz. ((**a**): calcite particulate; (**b**): quartz particulate; (c): distribution form of quartz, feldspar, and calcite; (**d**): distribution form of quartz, and smithsonite. Viewed under an optical microscope).

3.2. Flotation Experiments

It is well accepted that surface sulfidation of the mineral is an important prerequisite for the flotation of zinc-oxide minerals by sulfidation-xanthate flotation or sulfidation-amination flotation. In practice, Na_2S, as the sulphidizing agent was intensively used because of its low price, wide availability and ease of operation in most flotation systems, which could significantly improve the surface hydrophobicity of lead zinc oxide minerals [20,21]. It could increase the alkalinity of the solution and also act as a pH regulator [22]. Thus, sodium sulfide was selected as the sulphidizing agent for this lead-zinc oxide ore flotation.

Sodium carbonate was frequently employed as a pH adjuster and dispersant in flotation aqueous system [13,23]. In addition, the use of sodium carbonate with sodium silicate showed synergistic effects and improved the depression of complex calcium gangue minerals [24–26]. Meanwhile, sodium carbonate could react with Ca^{2+} to eliminate the adverse effects of Ca^{2+} by charge reversal [27]. Sodium silicate, contained a variety of compounds made of SiO_2 and Na_2O in varying ratios. It was the major depressant employed in the selective inhibition of gangue minerals such as quartz, talc, fluorite, dolomite, and calcite [28,29]. At present, the inhibition effect of sodium silicate on gangue minerals is still undesirable, and the depressant that contains -COOH, $-SiO_3$, $-PO_4$, and -OH groups exhibits many merits [30,31]. Therefore, an agent containing multiple carboxyl and hydroxyl functional groups (DCCH) was selected as an organic inhibitor in this paper. The effect of a mixed inhibitor (DCCH+ sodium silicate) on flotation was studied for the first time through flotation condition tests.

The most common flotation collectors used commercially for the treatment of zinc oxide minerals are cationic collectors and sulfhydryl collectors, such as amine and xanthate [32]. As is well known, slime causes significant negative effects on the flotation with amine, and xanthate exhibited a relatively poor collecting capacity for the zinc oxide minerals. The mixed anionic and cationic collectors were, therefore, used for the recovery of zinc oxide minerals in the study [33].

3.2.1. Flotation Conditional Experiments

A flow sheet consisting of "lead preferred flotation-zinc flotation" was employed (Figure 7), and the effects of the main experimental factors on the lead flotation were investigated in detail (as displayed in Figure 8). Five grinding settings, accounting for 70–90% with particle size −74µm were selected as independent variables to explore their effects on lead recovery and grade. The flotation recovery of lead increased with an increase in grinding fineness and then tended to become stable after accounting for 85% with particle size −74µm, while lead grade continued to decline. Therefore, the recommended grinding fineness was accounting for 85% with particle size −74µm (Figure 8a).

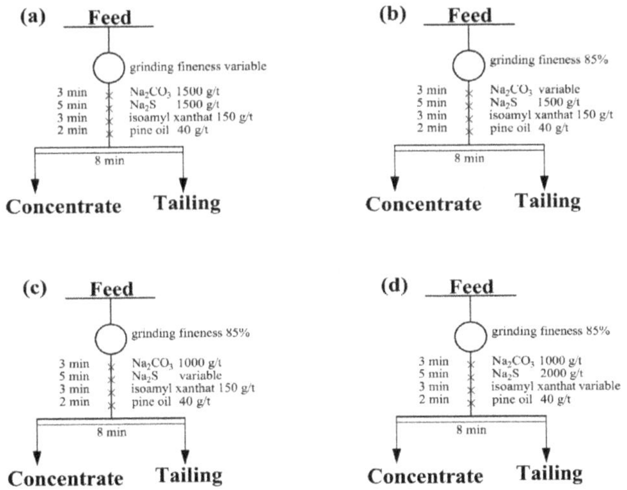

Figure 7. Flowsheet of the lead flotation conditional experiment. ((**a**) grinding fineness; (**b**) Na_2CO_3 dosage; (**c**) Na_2S dosage; (**d**) isoamyl xanthate dosage).

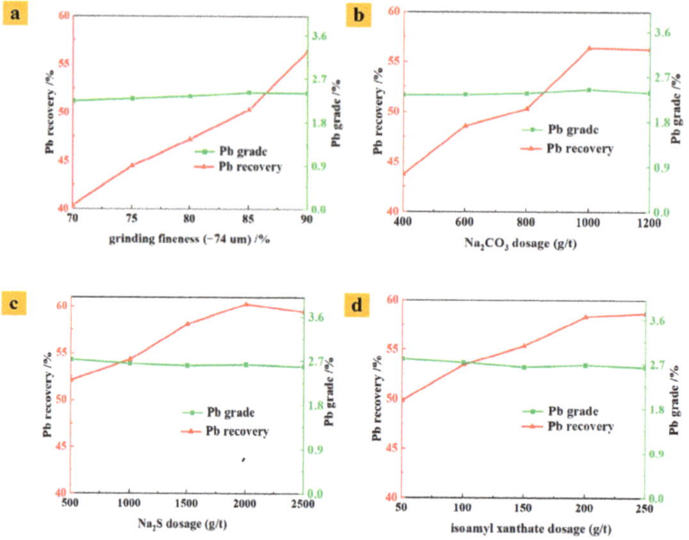

Figure 8. Effect of main experimental factors on the lead flotation. ((**a**): grinding fineness; (**b**): Na_2CO_3 dosage; (**c**): Na_2S dosage; (**d**): isoamyl xanthate dosage).

The impact of sodium carbonate on lead flotation was investigated (Figure 8b). Sodium carbonate had little effect on lead grade, but the flotation recovery increased considerably with increasing Na_2CO_3 dosage and reached the maximal value when the Na_2CO_3 dosage was 1000 g/t. Nevertheless, with a further increase in the Na_2CO_3 dosage, the recovery of lead failed to increase. So, 1000 g/t Na_2CO_3 was selected as the optimal dosage in the subsequent experiments.

The effect of the Na_2S dosage on lead flotation is shown in Figure 8c. The dosage of Na_2S exerted a positive role in lead flotation. When the dosage of Na_2S was 2000 g/t, the Pb recovery and grade in the concentrate achieved 59.32% and 2.63%. Simultaneously, the flotation recoveries decreased, and the Pb grade showed a downtrend when the Na_2S dosage exceeded this threshold value. It was possible that excess sodium sulfide depressed the lead minerals, especially for galena. Thus, the recommended Na_2S dosage was 2000 g/t.

The flotation recovery of lead significantly increased with the increase of isoamyl xanthate dosage, and the Pb grade presented the opposite trend. The Pb recovery was about 58% with the addition of 200 g/t isoamyl xanthate (Figure 8d). With a further increase in isoamyl xanthate dosage, the variation in Pb recovery was feeble. Thus, 200 g/t isoamyl xanthate was selected for the lead flotation.

Figure 9 shows the fluctuation amplitudes of the lead recovery rate and grade corresponding to grinding fineness, sodium carbonate, sodium sulfide, and isoamyl xanthate dosage, respectively. As can be seen, the lead recovery rate fluctuated greatly, and the lead grade fluctuated feebly. Among these factors, the most influential factor on flotation indexes was grinding fineness. The smallest influential factor was sodium sulfide dosage followed by isoamyl xanthate and sodium carbonate dosage.

The obtained tailing of lead flotation was used for the feed of zinc flotation, and the flow sheet is shown in Figure 10. A roughing process was performed for each dose investigation. The flotation time of sodium sulfide, DCCH + sodium silicate, cationic collector, and anionic collector was 5, 3, and 3 min, respectively, and froth collection was 8 min. The effects of the main experimental factors on the zinc flotation were investigated in detail (as displayed in Figure 11). From Figure 11a, the Zn recovery increased as Na_2S dosage increased from 6000 to 14,000 g/t, and the Zn grade in the concentrate ranged from 25.83%~26.63%. With a further increase in the Na_2S dosage to 14,000 g/t, the Zn grade decreased to some extent, and the Zn recovery had little change. Thus, 12,000 g/t Na_2S was selected as the optimal dosage in the subsequent experiments of zinc flotation.

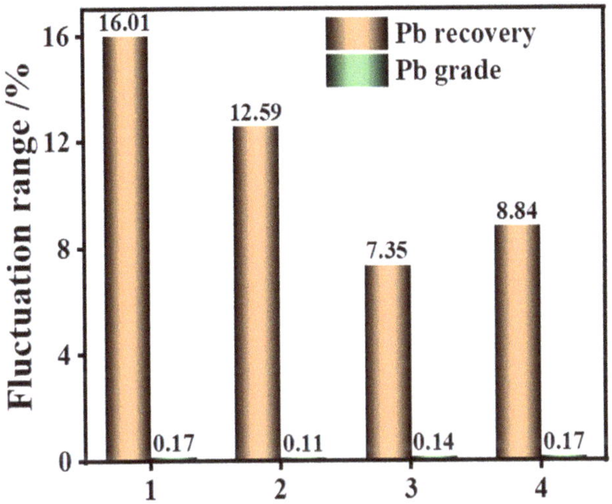

Figure 9. Influence of various experimental factors on lead flotation indexes. (1: grinding fineness; 2: Na_2CO_3 dosage; 3: Na_2S dosage; 4: isoamyl xanthate dosage).

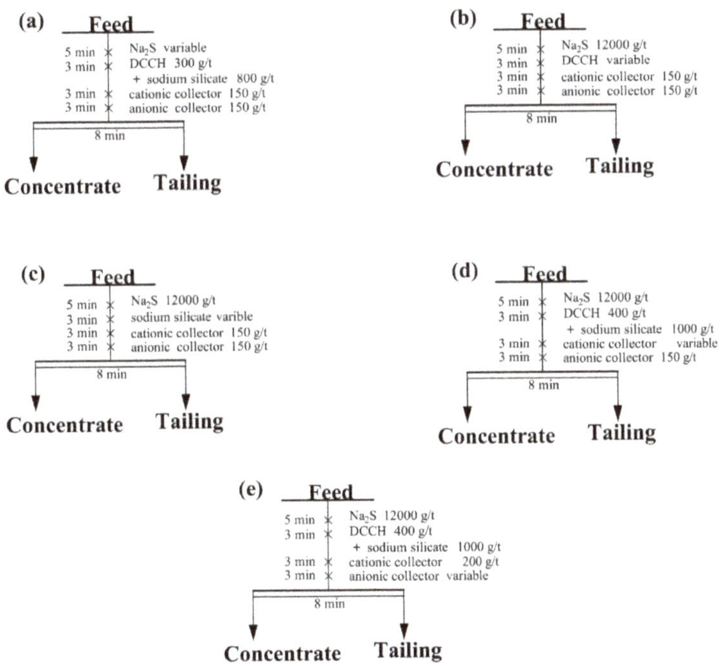

Figure 10. Flowsheet of the zinc flotation conditional experiment. ((**a**) Na_2S dosage; (**b**) DCCH dosage; (**c**) sodium silicate dosage; (**d**) cationic collector dosage; (**e**) anionic collector dosage).

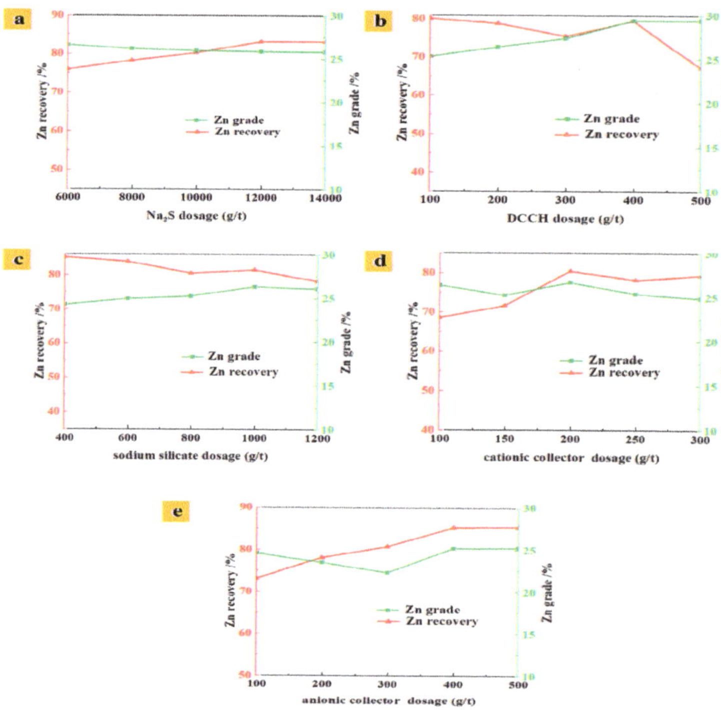

Figure 11. Effect of the agent dosage on the zinc flotation. ((**a**): Na_2S dosage, (**b**): DCCH dosage, (**c**): sodium silicate dosage, (**d**): cationic collector dosage, (**e**): anionic collector dosage).

The addition of suitable depressants was inevitable for the selective flotation of smithsonite from the gangue minerals because the smithsonite exhibited similar floatability as the carbonate gangue minerals (such as calcite, dolomite, and siderite) in the absence of depressants. Figure 11b,c shows the effect of DCCH and sodium silicate dosage on zinc flotation. The zinc grade in the concentrate increased from 25.33% to 29.44% when the DCCH increased from 100 to 500 g/t, and the zinc recovery presented a downtrend. Notably, the zinc recovery decreased to 67.33% with the addition of 500 g/t DCCH (Figure 11b). This indicated that excess DCCH could depress the smithsonite flotation considerably. Thus, the recommended dosage of DCCH was 400 g/t. The dosage of sodium silicate played an important role in the zinc flotation and showed a similar effect with DCCH on the zinc flotation indexes (Figure 11c). A sodium silicate dosage of 1000 g/t was reasonable; the Zn grade and recovery of concentrate were 26.35% and 81.58%, respectively. Trial test results of mixed depressants indicated that the combined use of two depressants facilitated the improvement of the zinc grade. When the mixed depressants (DCCH 400 g/t + sodium silicate 1000 g/t) were added, the grade and recovery arrived at 26.59% and 82.21%, displaying the superior selective flotation of zinc minerals with the aid of the desirable depressing effects of the mixed depressants on the carbonate gangue minerals.

As shown in Figure 10d, the flotation recoveries of zinc increased with the increase in the cationic collector dosage. Specifically, the Zn recovery sharply increased to 80.57%, and the Zn grade was 26.74% as the cationic collector dosage increased to 200 g/t. With a further increase in the cationic collector dosage, the Zn recovery changed feebly, and the Zn grade decreased by 1.32%. Thus, the optimum dosage of cationic collectors was determined as 200 g/t, which exhibited a superior collecting capacity for the zinc oxide minerals with the aid of an anionic collector (300 g/t). The effect of anionic collector dosage on zinc flotation was also investigated. As shown in Figure 10e, the recovery of zinc increased with the increase the dosage of the anionic collector, which reached 85.33%. The Zn grade was 25.22% in the presence of 400 g/t of an anionic collector. Thus, mixed collectors could effectively collect the zinc oxide minerals.

Figure 12 compares fluctuation amplitudes of the zinc recovery rate and grade corresponding to each factor. It could be concluded that DCCH exhibited the largest fluctuation ranges in both recovery and grade, followed by the anionic collector, cationic collector, sodium silicate, and sodium sulfide dosage.

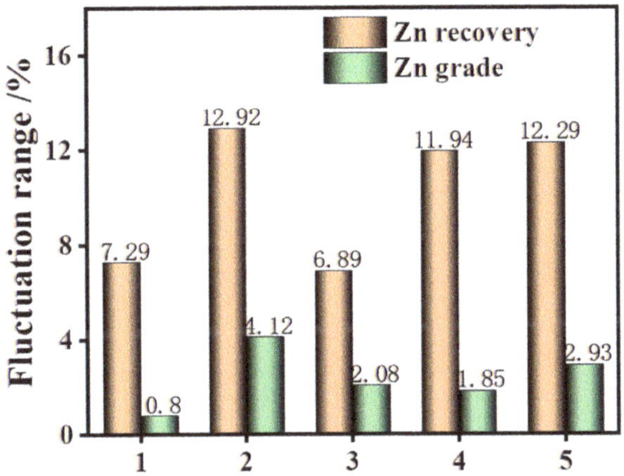

Figure 12. Influence of various experimental factors on zinc flotation indexes. (1: Na_2S dosage, 2: DCCH dosage, 3: sodium silicate dosage, 4: cationic collector dosage, 5: anionic collector dosage).

3.2.2. Flotation Flowchart Experiment

According to the results of the flotation conditional experiment, the flotation flowchart experiment was investigated (as shown in Figure 13) and the results are shown in Table 4. It is worth noting that an all-open flotation process consisting of "two-times lead rougher-one-time lead cleaner-two-times zinc-one-time zinc cleaner" was finally employed for the beneficiation of this ore. On the one hand, the present reagent regime in the roughing processing could effectively recover the lead and zinc minerals, and the Zn grade in the concentrates met the needs of subsequent metallurgical operations. On the other hand, the closed-circuit process of lead or zinc seriously deteriorated the flotation indexes based on the multitude of verification tests. The main reason was that the return of the middles resulted in the accumulation of the slimes, resulting in the non-selective flotation of target minerals. From Table 4, the Pb grade in the lead rough concentrate was 2.83% and the Pb and recovery was 57.56%. Meanwhile, the loss rate of zinc was only 11.77%. The Zn grade in the Zn concentrate reached 28.64% with a recovery of 83.45%. The Zn grade in the final tailings was only 0.59%. Thus, the presence of a flotation flowchart and reagent regime resulted in the effective beneficiation of this low-grade oxidized lead and zinc ore, which could provide important technical support for the processing of similar ores.

Table 4. The results of the opened-circuit flotation experiment of oxidized lead and zinc ore.

Product	Yield/%	Grade/%		Recovery/%	
		Pb	Zn	Pb	Zn
Pb rougher concentrate	17.29	2.83	5.14	57.56	11.77
Zn concentrate	22.00	0.23	28.64	5.59	83.45
Tailing	60.71	0.51	0.59	36.49	4.78
Total	100.00	0.85	7.55	100.00	100.00

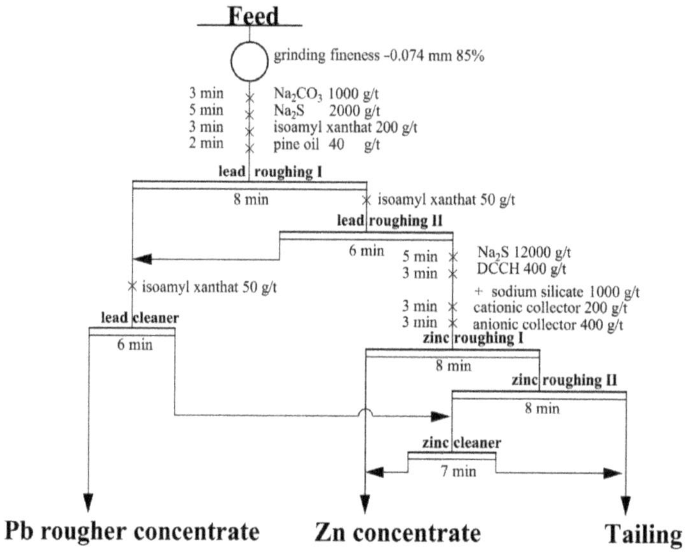

Figure 13. The flowchart of the opened-circuit flotation experiment.

4. Conclusions

Flotation optimization of a low-grade oxidized lead and zinc ore from the Lanping mine was conducted on the basis of process mineralogy characteristics. The following conclusions could be drawn.

The Pb and Zn grades in the raw ore were 0.84% and 7.42%. Smithsonite and cerussite were the main valuable metallic minerals; calcite ($CaCO_3$), siderite ($FeCO_3$), and quartz (SiO_2) were the main gangue minerals. A small amount of galena and sphalerite were concomitant in the ore. The low content of lead and zinc and the complex co-occurrence relationships between minerals, especially iron and zinc, would bring a challenge to the efficient separation of the ore.

A flow sheet consisting of "lead preferred flotation-zinc flotation" was employed. The best performance for lead flotation was obtained at a grinding fineness of accounting for 85% with particle size −74 μm, a Na_2CO_3 dosage of 1000 g/t, a Na_2S dosage of 2000 g/t, an isoamyl xanthate dosage of 200 g/t and pine oil of 40 g/t. The recommended agent dosages for zinc roughing were as follows: the Na_2S dosage was 12,000 g/t, the mixed depressants dosage was 400 g/t DCCH and 1000 g/t sodium silicate; the mixed collector dosages were 200 g/t cationic collector and 400 g/t anionic collector. In addition, the influence order of experimental factors on lead flotation was as follows: grinding fineness > sodium carbonate > isoamyl xanthate > sodium sulfide. The influence order of experimental factors on zinc flotation was as follows: DCCH, > anionic collector > cationic collector > sodium silicate > sodium sulfide. An all-open flotation process consisting of "two-times lead rougher-one-time lead cleaner-two-times zinc-one-time zinc cleaner" was finally employed for the beneficiation of this ore. The Pb grade in the lead rough concentrate was 2.83%, and the Pb and recovery was 57.56%. The Zn grade in the Zn concentrate reached 28.64% with a recovery of 83.45%. The Zn grade in the final tailings was only 0.59%. Thus, the presence of a flotation flowchart and reagent regime effectively beneficiates this ore, which could provide important technical support for the processing of similar ores. However, attention must be paid to the optimization of the process flow and reagent systems for industrial applications, such as the reduction of sodium sulfide dosage and the increase in the Pb concentrate grade.

Author Contributions: Conceptualization, Writing-review and editing, A.Y.; Investigation, Data curation, Z.D.; Conceptualization, Methodology, Visualization, Investigation, J.Y.; Conceptualization and review, Q.F.; Conceptualization, writing review and editing, S.B.; Supervision, S.W. All authors have read and agreed to the published version of the manuscript.

Funding: This work was supported by the National Natural Science Foundation of China (Grant No.52164021); the Natural Science Foundation of Yunnan Province (Grant No. 2019FB078); and the Special Funds for this research project (Grant No. CCC21321119A and Grant No. CA22369M062A).

Data Availability Statement: All data generated during the current study are available in full with in this article.

Conflicts of Interest: The authors declare no conflict of interest.

References

1. Moradi, S.; Monhemius, A.J. Mixed sulphide–oxide lead and zinc ores: Problems and solutions. *Miner. Eng.* **2011**, *24*, 1062–1076. [CrossRef]
2. Saidi, M.; Kadkhodayan, H. Experimental and theoretical evaluation of zinc recovery from zinc oxide ore: Process optimization and simulation using Aspen Plus software. *Int. J. Chem. React. Eng.* **2020**, *18*, 103772. [CrossRef]
3. Ahmed, I.M.; Nayl, A.A.; Daoud, J.A. Leaching and Recovery of Zinc and Copper from Brass Slag by Sulfuric Acid. *J. Saudi Chem. Soc.* **2016**, *20*, S280–S285. [CrossRef]
4. Brough, C.P.; Warrender, R.; Bowell, R.J.; Barnes, A.; Parbhakar-Fox, A. The process mineralogy of mine wastes. *Miner. Eng.* **2013**, *52*, 125–135. [CrossRef]
5. Lotter, N.O.; Baum, W.; Reeves, S.; Arrué̇d, C.; Bradshaw, D.J. The business value of best practice process mineralogy. *Miner. Eng.* **2018**, *116*, 226–238. [CrossRef]
6. Cabri, L.J.; Beattie, M.; Rudashevsky, N.S.; Rudashevsky, V.N. Process mineralogy of Au, Pd and Pt ores from the Skaergaard intrusion, Greenland, using new technology. *Miner. Eng.* **2005**, *18*, 887–897. [CrossRef]
7. Bastien, D.; Deyan, M.; Hassan, B.; Stoyan, G. Mineralogical study of electrum grain size, shape and mineral chemistry in process streams from the Krumovgrad mine, Bulgaria. *Miner. Eng.* **2023**, *198*, 108080.
8. Bahrami, A.; Abdollahi, M.; Mirmohammadi, M.; Kazemi, F.; Shokrzadeh, M. A Process mineralogy approach to study the efficiency of milling of molybdenite circuit processing. *Sci. Rep.* **2020**, *10*, 21211. [CrossRef]
9. Li, H.; Wu, X.; Wang, M.; Wang, J.; Wu, S.; Yao, X.; Li, L. Separation of elemental sulfur from zinc concentrate direct leaching residue by vacuum distillation. *Sep. Purif. Technol.* **2014**, *138*, 41–46. [CrossRef]
10. Muravyov, M. Zinc leaching from sulfidic concentrate in two-step biohydrometallurgical process. *J. Biotechnol.* **2019**, *305*, S51. [CrossRef]
11. Tsogtkhankhai, D.; Mamyachenkov, S.V.; Anisimova, O.S.; Naboichenko, S.S. Thermodynamics of reactions during nitric acid leaching of minerals of a copper concentrate. *Russ. J. Non-Ferr. Met.* **2011**, *52*, 135–139. [CrossRef]
12. Chaudhury, G.R.; Sukla, L.B.; Das, R.P. Kinetics of biochemical leaching of sphalerite concentrate. *Metall. Trans. B* **1985**, *16*, 667–670. [CrossRef]
13. Wang, L.; Gao, Z.; Tang, H.; Wang, L.; Han, H.; Sun, W.; Qu, Y.; Yang, Y. Copper recovery from copper slags through flotation enhanced by sodium carbonate synergistic mechanical activation. *J. Environ. Chem. Eng.* **2022**, *10*, 107671.
14. Hosseini, S.H.; Forssberg, E. Physicochemical studies of smithsonite flotation using mixed anionic/cationic collector. *Miner. Eng.* **2007**, *20*, 621–624. [CrossRef]
15. Wang, L.; Lyu, W.; Zhou, W.; Zhang, H. The role of sodium phytate in the flotation separation of smithsonite from calcite. *Miner. Eng.* **2022**, *23*, 107775. [CrossRef]
16. Wen, S.M.; Feng, Q.C. Formation of zinc sulfide species on smithsonite surfaces and its response to flotation performance. *J. Alloys Compd.* **2017**, *709*, 602–608.
17. Li, Y.; Wang, J.K.; Wei, C.; Liu, C.X.; Jiang, J.B.; Wang, F.C. Sulfidation roasting of low grade lead–zinc oxide ore with elemental sulfur. *Miner. Eng.* **2010**, *23*, 563–566. [CrossRef]
18. Da, J.; Ruofan, S.; Guoyong, W.; Jiushuai, D.; Xi, Z. Flotation separation of fluorite and calcite using anhydrous glucose and aluminum sulfate as a combined depressant. *Appl. Surf. Sci.* **2023**, *624*, 157089.
19. Wang, M.T.; Zhang, G.F.; Zhao, L.; Chen, Y.F.; Liu, D.Z.; Li, C.B. Application of eco-friendly tetrasodium iminodisuccinate for separation of smithsonite from zinc ions activated quartz. *Miner. Eng.* **2022**, *181*, 107545. [CrossRef]
20. Salum, M.J.G.; De Araujo, A.C.; Peres, A.E.C. The role of sodium sulphide in amine flotation of silicate zinc minerals. *Miner. Eng.* **1992**, *5*, 411–419. [CrossRef]
21. Liu, C.; Zhang, W.C.; Song, S.X.; Li, H.Q.; Jiao, X.K. A novel insight of the effect of sodium chloride on the sulfidization flotation of cerussite. *Powder Technol.* **2019**, *344*, 103–107. [CrossRef]
22. Atani, S.E.; Tekir, U.; Sebitla, L.D. Effects of additives in froth flotation of silicate zinc ore; a study by zeta potential measurement and infrared spectroscopy. *Chiang Mai J. Sci.* **2007**, *34*, 191–200.

23. Wang, Y.H.; Sun, D.X.; Wang, L.G.; Zhou, Y.L. Effects of sodium tripolyphosphate and sodium carbonate on the selective flocculation of diasporic-bauxite in the presence of calcium and magnesium ions. *Miner. Eng.* **2011**, *24*, 1031–1037.
24. Filippov, L.O.; Foucaud, Y.; Filippova, I.V.; Badawi, M. New reagent formulations for selective flotation of scheelite from a skarn ore with complex calcium minerals gangue. *Miner. Eng.* **2018**, *123*, 85–94. [CrossRef]
25. Cheng, L.; Feng, Q.M.; Zhang, G.F.; Chen, W.; Chen, Y.F. Effect of depressants in the selective flotation of scheelite and calcite using oxidized paraffin soap as collector. *Int. J. Miner. Process.* **2016**, *157*, 210–215.
26. Kupka, N.; Martin, R. Role of sodium carbonate in scheelite flotation—A multi-faceted reagent. *Miner. Eng.* **2018**, *129*, 120–128. [CrossRef]
27. Chen, Y.; Zhang, G.; Wang, M.; Shi, Q.; Liu, D.; Li, Q. Utilization of sodium carbonate to eliminate the adverse effect of Ca^{2+} on smithsonite sulphidisation flotation. *Miner. Eng.* **2019**, *132*, 121–125. [CrossRef]
28. Kupka, N.; Rudolph, M. Froth flotation of scheelite—A review. *Int. J. Min. Sci. Technol.* **2018**, *28*, 373–384. [CrossRef]
29. Marion, C.; Li, R.; Waters, K.E. A review of reagents applied to rare-earth mineral flotation. *Adv. Colloid Interface Sci.* **2020**, *279*, 102142. [CrossRef] [PubMed]
30. Qin, W.Q.; Hu, J.J.; Zhu, H.L.; Jiao, F.; Jia, W.H.; Han, J.W.; Chen, W.Q. Effect of depressants on flotation separation of magnesite from dolomite and calcite. *Int. J. Min. Sci. Technol.* **2023**, *33*, 83–91. (In Chinese) [CrossRef]
31. Qin, W.Q.; Hu, J.J.; Zhu, H.L.; Jiao, F.; Pan, Z.C.; Jia, W.H.; Han, J.W.; Chen, C. Selective inhibition mechanism of PBTCA on flotation separation of magnesite from calcite. *Colloids Surf. A* **2021**, *630*, 127597. [CrossRef]
32. Irannajad, M.; Ejtemaei, M.; Gharabaghi, M. The effect of reagents on selective flotation of smithsonite–calcite–quartz. *Miner. Eng.* **2009**, *22*, 766–771. [CrossRef]
33. Ejtemaei, M.; Irannajad, M.; Gharabaghi, M. Influence of important factors on flotation of zinc oxide mineral using cationic, anionic and mixed (cationic/anionic) collectors. *Miner. Eng.* **2011**, *24*, 1402–1408. [CrossRef]

minerals

MDPI

Article

Utilization of PMA-EDTC as a Novel Macromolecular Depressant for Galena in the Flotation Separation of Chalcopyrite

Hong Zeng [1,2,*], Yangge Zhu [2], Chuanyao Sun [2], Zhiqiang Zhao [2], Guiye Wu [2], Chongjun Liu [2], Tong Lu [2] and Xingrong Zhang [3]

1 School of Civil and Resources Engineering, University of Science and Technology Beijing, Beijing 100083, China
2 State Key Laboratory of Mineral Processing Science and Technology, BGRIMM Technology Group, Beijing 102600, China
3 College of Chemical Engineering, Eco-Chemical Engineering Cooperative Innovation, Center of Shandong, Qingdao University of Science & Technology, Qingdao 266042, China
* Correspondence: zenghong@bgrimm.com

Abstract: To address the issue of mediocre separation efficiency of depressants in the copper-lead separation process, this article synthesized a macromolecular organic depressant, polymaleic anhydride-ethylenediaminetetraacetic acid (PMA–EDTC), using a polycarboxylic macromolecule as the backbone and also introducing –N–(C=S)–S– as the solidophilic group and employed as a galena depressant. The structure of PMA–EDTC was characterized using Fourier transform infrared (FT-IR). The effect of PMA–EDTC on the floatability of galena and chalcopyrite was investigated through micro-flotation and Contact angle measurements. The experimental results demonstrated that PMA–EDTC exhibited selectivity inhibition towards galena rather than chalcopyrite across a wide pH range. At a dose of 8 mg/L, there was effective separation between galena and chalcopyrite with a separation coefficient of 24.17, effectively altering the floatability of galena while having little impact on the floatability of chalcopyrite. The selective inhibition behavior and adsorption mechanism of PMA–EDTC on galena and chalcopyrite were investigated using FT-IR, Zeta potential, and X-ray photoelectron spectroscopy (XPS). FT-IR and Zeta potential studies indicated that PMA–EDTC formed chemical adsorption on the surface of galena. XPS confirmed the model of chemical adsorption of PMA–EDTC on lead atoms in the galena surface. The results indicate that PMA–EDTC adsorbs on the surface of galena via its –(C=S)–S–group, forming a hydrophilic complex and achieving selective depression of lead and the cleaning flotation of copper.

Keywords: galena; chalcopyrite; depressant; flotation separation; adsorption mechanism

1. Introduction

Copper and lead are widely employed in various fields such as national defense technology, aerospace engineering, the automotive industry, construction, and electronics [1,2]. With the continuous increase in demand for copper and lead, easily exploitable rich ores are gradually depleted, leading to an increasing dependence on complex low-grade ores [1,3,4]. Concurrently, with the rapid progress of technology, driven by the requirements of new functional materials, the quality standards of copper and lead concentrate are becoming more and more stringent, which puts forward higher requirements for the flotation process and flotation reagents of complex copper and lead ore separation [5–7].

In the traditional copper-lead separation process, cyanide and its combination reagents are usually used to "reduce galena and float copper" and dichromate is used to "reduce lead and float copper". The cyanide and potassium dichromate used are highly toxic agents [8–10]. Although these traditional inorganic inhibitors have good selectivity, their use is limited due to their high toxicity and severe environmental pollution [11–13].

Compared with the toxicity of inorganic inhibitors, small molecular organic inhibitors have the advantages of low toxicity and good selectivity. In recent years, organic inhibitors have attracted more and more attention and become a research hotspot in the field of flotation [14]. According to recent studies, the application of DPS, pseudo glycolythiourea acid, disodium carboxymethyltrithiocarbonate, thioglycolic acid, 2,3-disulfanylbutanedioic acid [15–18] in the separation of sulfide minerals demonstrates that it can be used for flotation separation of sulfide ore. The depressing activity of new organic depressant ACS to chalcopyrite and galena has been studied via artificial mixed ore and actual ore tests. The results show that ACS can depress galena well and cannot depress chalcopyrite [19]. A small reagent, O,O-bis(2,3-dihydroxypropyl) dithiophosphate (DHDTP), was investigated as a depressant on the depression of chalcopyrite and galena when ammonium dibutyl dithiophosphate (DDTP) was used as the collector in flotation tests. The flotation tests show that DHDTP has a strong depression on galena and a slight depression on chalcopyrite in a wide pH range. The Zeta potential and adsorption measurements show that DHDTP has a stronger adsorption capacity to galena than chalcopyrite [20]. Although much effort has been focused on the research of small organic inhibitors, due to their high price or poor performance in pilot scale tests, it is still a change task to apply these compounds on a commercial scale.

Natural high molecular weight organic inhibitors such as lignin, humic acid, polysaccharides, and synthetic high molecular weight organic inhibitors such as sodium polyacrylate and polyacrylamide derivatives have also been applied and studied due to their stronger hydrophilicity and lower drug consumption [21–24]. Recent research finds that synthetic large-molecule organic depressants exhibit remarkable advantages. These advantages include a diverse array of types, varied sources of raw materials, and the capacity for tailoring functional groups and molecular weights based on specific requirements [25]. Moudgil et al. [26] found that nonionic polyacrylamide possesses a certain inhibitory effect on coal. Nonionic polyacrylamide will adsorb on the surface of coal particles and render their surface hydrophilic. Boulton et al. [27] introduced several functional groups, such as carboxyl, sulfonic acid, hydroxyl, or thiourea, onto low molecular weight polyacrylamide structures to produce various polyacrylamide derivatives with unique properties. These derivatives were then utilized as pyrite inhibitors. Zhang et al. [28] explored the inhibitory effect of hydroxame polyacrylamide on pyrite and studied the mechanism of action of hydroxame polyacrylamide on the surface of pyrite. Consequently, the molecular design and modification of large-molecule organic depressants offer opportunities for enhancing their selectivity and inhibitory efficiency [29]. This has the potential to overcome the limitations of existing high molecular weight organic inhibitors, while also helping to address the environmental concerns of conventional inhibitors.

In this paper, an organic macromolecular galena inhibitor based on a polymaleic anhydride skeleton was designed, synthesized, and applied to the separation of copper and lead. The effectiveness of the developed organic macromolecular inhibitor in promoting the flotation separation of copper lead sulfide minerals was investigated through flotation experiments of single minerals and artificially mixed minerals. Zeta potential measurement, FT-IR spectroscopy, and X-ray diffraction (XRD) analysis techniques have been used to explore potential inhibition mechanisms, predict interaction models, and provide a theoretical basis for designing organic macromolecular inhibitors suitable for galena.

2. Materials and Methods

2.1. Minerals and Reagents

Galena and chalcopyrite samples with relatively high grades were procured from a mining site located in Hubei Province, China. These samples were crushed and finely ground in a ceramic ball mill. Subsequently, particle sizes ranging from −74 μm to +38 μm were dry-sieved in micro-flotation experiments. The X-ray diffraction (XRD) spectra and multielement analysis of both chalcopyrite and galena are presented in Figure 1 and Table 1, respectively.

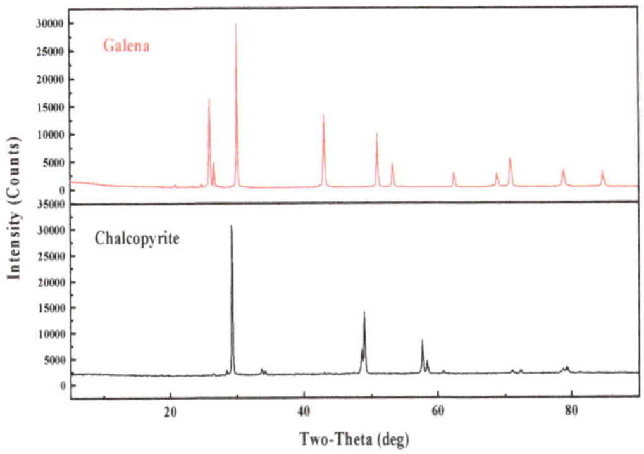

Figure 1. X-ray diffraction (XRD) results of the galena and chalcopyrite samples.

Table 1. Multi-element assay of the galena and chalcopyrite samples.

Sample	P	Zn	Pb	SiO_2	Al_2O_3	CaO	MgO	Cu	Tfe	S
Chalcopyrite	0.13	0.075	0.016	<0.01	0.040	0.11	0.078	33.95	30.84	34.54
Galena	0.0021	<0.01	84.99	<0.01	0.047	0.032	0.013	0.031	0.067	12.11

Based on X-ray diffraction (XRD) and multi-elements assay, the purity of the galena and chalcopyrite minerals sample is notably high, with galena purity at 98.14% and chalcopyrite purity at 98.23%.

The collector used in the flotation experiments, O-(2-methylpropyl) N-ethylcarbamothi oate (Z-200), was sourced from BGRIMM Chemical Technology Co., Ltd. (Cangzhou, China), while the frother employed, methyl isobutyl carbinol (MIBC), was purchased from McLean Biochemical Technology Co., Ltd. (Langfang, China). Sodium hydroxide (NaOH) and hydrochloric acid (HCl), utilized as pH adjusters, were purchased from Sinopharm Chemical Reagent Co., Ltd. (Shanghai, China). The synthetic raw materials including maleic anhydride (MA), ethylenediamine, and carbon disulfide were also purchased from China National Pharmaceutical Group Chemical Reagent Co., Ltd. (Shanghai, China). All chemical reagents used above are analytical grade. Experiments were conducted in a deionized water environment.

2.2. Micro-Flotation Experiments

The mico-flotation experiments with pure minerals were conducted using a 30 mL XFGII flotation cell. Prior to micro-flotation, 2.0 g of pure mineral sample was placed into a beaker containing 30 mL of deionized water, and subjected to 5 min of ultrasonic cleaning to remove surface oxides from the minerals. After 10 min of settling, the suspension was decanted, and the bottom layer of pure minerals was transferred to the flotation cell. Subsequently, 30 mL of deionized water was added, and the required flotation reagents such as NaOH, HCl, PMA–EDTC, Z-200, and MIBC were sequentially introduced during the stirring process in the slurry. Each reagent addition was followed by a 2-min stirring period. Finally, the foam product and tailings were filtered, dried, and weighed. The recovery rates of each mineral were calculated using Formula (1). Each experiment was repeated three times to obtain an average recovery.

In the micro-flotation experiments with artificially mixed minerals, a mixture containing 1.2 g of galena and 0.8 g of chalcopyrite was used as the feed. The flotation procedure was consistent with the single mineral flotation tests. The copper and lead grades in the

Table 2. PMA–EDTC gel permeation chromatography (GPC) test results.

	Mp	Mn	Mw	Mz	Mz + 1	Mv	PD
PMA–EDTC	2197	1753	2067	2402	2738	2018	1.18

The FT-IR spectra of polymers PMA and PMA–EDTC are shown in Figure 5. C–H stretching vibration at 3056.80 cm^{-1}; C=O stretching vibration at 1719.71 cm^{-1}; C–H–O in-plane bending vibration of carboxylic acid [30] at 1413 cm^{-1}; C–O stretching vibration at 939.03 cm^{-1}; asymmetric stretching vibration of the epoxide ring at 823.14 cm^{-1} in the FT-IR spectrum of PMA; 1234.81 cm^{-1} is the skeleton vibration of the epoxide ring. In contrast, three new peaks appear in the IR spectra of PMA after the reaction, the peak at 1553.43 cm^{-1} is probably caused by N–H bending vibration, the peak at 1347.55 cm^{-1} is caused by C–N stretching vibration, the absorption peak at 1030.54 cm^{-1} is caused by C=S stretching vibration, the peak at 3412.32 cm^{-1} The peak at 3412.32 cm^{-1} is caused by the –NH_2 stretching vibration, the correctness of its structure was verified by FT-IR spectra [31].

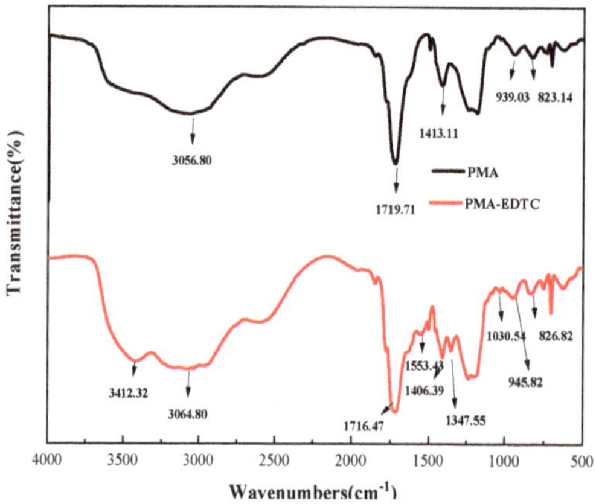

Figure 5. Infrared spectra of PMA and PMA–EDTC.

3.2. Micro-Flotation Experiments

In the single mineral flotation tests, the inhibitory performance of PMA–EDTC was investigated by varying the inhibitor dosage and the slurry pH, as depicted in Figure 6.

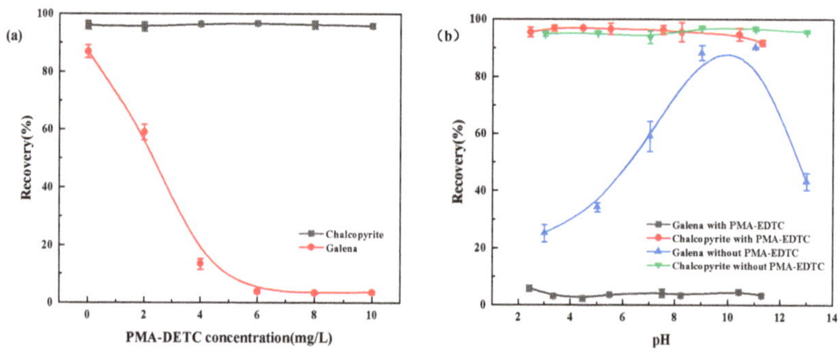

Figure 6. The relationship between recovery and (**a**) PMA–EDTC concentration and (**b**) pH.

As shown in the graph, at a slurry pH around 8.5, with a frother MIBC dosage of 10.0 mg/L and a collector Z-200 dosage of 10.0 mg/L, the recovery rate of galena (lead mineral) exhibited a rapid decline with increasing PMA–EDTC dosage, while the recovery rate of chalcopyrite (brass mineral) remained relatively stable, staying above 95%. When the PMA–EDTC dosage reached 6.0 mg/L, the recovery rate of galena dropped to 4.74%, presenting a difference of around 90% compared to chalcopyrite, underscoring the effective inhibitory role of PMA–EDTC against galena. Further increasing the PMA–EDTC dosage did not significantly alter the recovery rate of galena, thereby confirming the optimal dosage of PMA–EDTC as 6.0 mg/L.

Subsequent investigation into the influence of slurry pH on the inhibitory performance of PMA–EDTC revealed that its lead inhibitory capability remained relatively consistent across a range of slurry pH values. In the pH range of 2.0 to 12.0, PMA–EDTC consistently exhibited a strong inhibitory effect on galena while maintaining minimal impact on the floatability of chalcopyrite. Building upon the foundation of single mineral flotation experiments, additional validation of PMA–EDTC copper-lead separation efficiency was conducted through artificial mixed mineral tests.

Table 3 presents the results of artificial mixed mineral flotation experiments under various PMA–EDTC reagent dosages. The amount of trapping agent Z-200 and blowing agent MIBC was the same as in the single mineral test, and the pH of the slurry was 8.5. In the absence of the inhibitory agent PMA–EDTC, the recovery rate of Pb in the concentrate is 97.10%, while the recovery rate of Cu is 99.89%. At this point, the copper-lead separation factor is 5.3. With increasing PMA–EDTC dosage, the recovery rate of galena gradually decreases, and the copper-lead separation factor increases. When the PMA–EDTC dosage reaches 8 mg/L, the recovery rate of Pb decreases to 0.56%, while the recovery rate of Cu is 76.82%. At this dosage, the copper-lead separation factor reaches 24.17. These findings indicate that under appropriate reagent dosage conditions, PMA–EDTC exhibits significant inhibitory effects on galena, making it an effective lead inhibitor for copper-lead separation.

Table 3. PMA–EDTC artificial mixed mineral test results.

PMA–EDTC Dosage/mg·L^{-1}	Product	Grade/%		Recovery/%		Separation Index *
		Pb	Cu	Pb	Cu	
	Concentrate	55.95	14.20	97.10	99.89	
0	Tailing	35.72	0.32	2.90	0.11	5.30
	Feed	55.05	13.58	100.00	100.00	
	Concentrate	36.39	19.04	43.56	93.90	
2	Tailing	80.05	2.10	56.44	6.10	4.46
	Feed	52.57	12.76	100.00	100.00	
	Concentrate	6.46	33.17	4.00	76.88	
4	Tailing	73.87	4.75	96.00	23.12	8.93
	Feed	52.12	13.92	100.00	100.00	
	Concentrate	3.02	33.46	1.72	76.49	
6	Tailing	75.66	4.52	98.28	23.51	13.61
	Feed	53.48	13.36	100.00	100.00	
	Concentrate	1.08	33.60	0.56	76.82	
8	Tailing	75.72	4.03	99.44	23.18	24.17
	Feed	54.49	12.44	100.00	100.00	
	Concentrate	2.16	33.18	0.40	25.48	
10	Tailing	61.42	11.08	99.60	74.52	9.22
	Feed	55.35	13.35	100.00	100.00	

* Separation index: Indicates the degree to which a unit separation operation or a separation process separates two substances.

3.3. Contact Angle Measurements

The contact angle is a quantitative measure of surface wetting, represented by the angle formed between a liquid or vapor interface and the solid surface. In general, a larger contact angle indicates a stronger hydrophobicity of the mineral surface, suggesting better floatability. Conversely, a smaller contact angle indicates a stronger hydrophilicity of the mineral surface, implying poorer floatability [32].

The images in Table 4 illustrate the contact angle variations of galena and chalcopyrite under natural pH conditions. Upon treatment with 20.0 mg/L PMA–EDTC, the contact angle on the surface of galena decreases from 79.8° to 60.0°, while the contact angle of chalcopyrite remains relatively unchanged. As the concentration of PMA–EDTC reaches 50.0 mg/L, the contact angle on the surface of the galena further decreases to 47.1°. These results indicate that PMA–EDTC can effectively modify the floatability of galena at appropriate concentrations, with minimal impact on the floatability of chalcopyrite. These findings are consistent with the results of flotation experiments, confirming the selective inhibitory effect of PMA–EDTC on the flotation of galena while having little effect on the floatability of chalcopyrite.

Table 4. Contact angle measurement results.

Without Any Treatment (Galena)	10.0 mg/L PMA−EDTC (Galena)	20.0 mg/L PMA−EDTC (Galena)	50.0 mg/L PMA−EDTC (Galena)

Without Any treatment (Chalcopyrite)	**10.0 mg/L PMA–EDTC (Chalcopyrite)**	**20.0 mg/L PMA–EDTC (Chalcopyrite)**	**50.0 mg/L PMA–EDTC (Chalcopyrite)**
		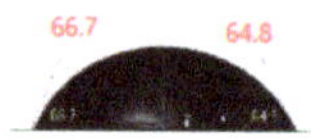	

3.4. *Zeta Potential Analysis*

Flotation experiments and contact angle measurements have demonstrated the effectiveness of PMA–EDTC in reducing the floatability of galena and consequently lowering its recovery rate. To further elucidate the inhibition mechanism of PMA–EDTC on galena, we conducted Zeta potential measurements on galena before and after treatment with PMA–EDTC. The results are presented in Figure 7.

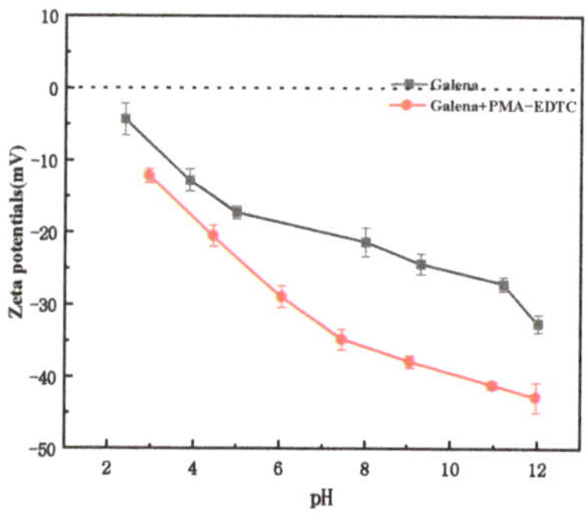

Figure 7. Effect of the pH on zeta potentals of galena before and after adding PMA-EDTC.

As depicted in Figure 7, it can be observed that the Zeta potential of galena remains consistently negative within the pH range of 2 to 12, aligning with prior reported observations. Upon the introduction of PMA–EDTC, a significant increase in the absolute value of the Zeta potential of galena is noted. This enhancement could be attributed to the presence of hydroxyl groups (–COOH) and dithiocarbamate groups (–NH–(C=S)–) within PMA–EDTC [33,34]. The chemical adsorption of PMA–EDTC onto the galena surface through its dithiocarbamate groups forms coordination complexes, while the carboxylate groups (–COO) are exposed to the aqueous phase, leading to the formation of a hydrophilic layer on the galena surface and subsequently resulting in a decrease in the Zeta potential of galena. These outcomes collectively indicate the adsorption of PMA–EDTC onto the galena surface.

3.5. FT-IR Analysis

In order to further unravel the adsorption mechanism of PMA–EDTC on galena surface, we compared FT-IR spectra of PMA–EDTC, PMA–EDTC/Pb^{2+} complexes, and galena before and after reaction with PMA–EDTC as shown in Figure 8.

Comparing the FT-IR spectrum of the PMA–EDTC/Pb^{2+} complex with that of PMA–EDTC (Figure 8a), it is evident that the C=S stretching vibration around 1234.67 cm^{-1} almost disappears in the complex spectrum. This disappearance suggests that C=S forms a metal coordination complex by interacting with Pb^{2+}. The positions of the $-NH_2$, N–H, and C–N absorption peaks in PMA–EDTC also undergo changes upon complex formation, likely due to rearrangements in the molecular structure [35,36]. Through the FT-IR spectra results, it is evident that PMA–EDTC further confirmed that a chemical reaction had occurred between (–NH–(C=S)–) group with galena, providing direct evidence of chemisorption of PMA–EDTC on the galena surface.

After treatment with PMA–EDTC (Figure 8b), the galena shows a distinct peak at 1711.31 cm^{-1}. By comparing the infrared spectra of PMA–EDTC, it is determined that this peak is due to the stretching vibration of C=O in PMA–EDTC. The peaks at 1528.00 cm^{-1} and 1323.98 cm^{-1} are respectively caused by the N–H bending vibration and C–N stretching vibration in PMA–EDTC. The absorption peaks at 1400.75 cm^{-1} and 1050.26 cm^{-1} are due to the internal stretching vibration of the carboxyl group in PMA–EDTC and the stretching vibration of C=S in –C(=S)S, respectively. The appearance of these peaks also indicates that the adsorption behavior of PMA–EDTC occurs on the surface of the galena.

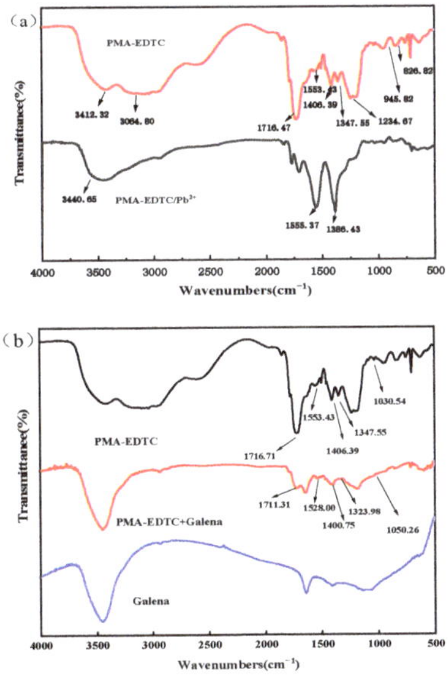

Figure 8. FT–IR spectra of (**a**) PMA–EDTC/Pb^{2+} complexes and (**b**) galena before and after PMA–EDTC treatment.

3.6. XPS Analysis

The chemical adsorption of PMA–EDTC on the galena surface, as well as its capability to form metal complexes with Pb^{2+} through reaction, has been confirmed through FT-IR spectra and zeta potential testing. To further elucidate the interaction mechanism between

PMA–EDTC and the galena surface, XPS experiments were conducted to analyze the surface characteristics of the galena before and after PMA–EDTC treatment. The XPS survey spectra of PMA–EDTC and the galena before and after treatment are presented in Figure 9, and the changes in surface concentrations of elements C, N, O, S, and Pb on galena surfaces due to PMA–EDTC treatment are summarized in Table 5. As observed from Figure 9, a newly emerged N 1s peak on the galena surface after treatment indicates the adsorption of PMA–EDTC. Notably from Table 5, the concentrations of O and C on the galena surface increase upon PMA–EDTC adsorption, while the concentrations of S and Pb decrease correspondingly. These variations are attributed to the adsorption of the inhibitor PMA–EDTC on the galena surface.

Table 5. Atomic concentration of elements with untreated and treated galena with PMA–EDTC and PMA–EDTC complexes.

Sample	Atomic Concentration of Elements (Atomic%)				
	C	O	N	S	Pb
PMA–EDTC	63.76	29.27	5.98	1.00	
Galena+PMA–EDTC	50.06	22.96	1.43	16.12	9.43
Galena	11.28	9.88		54.14	24.70
Δa	38.78	13.08	1.43	−36.02	−15.27

To reveal the surface atomic states of the minerals, high-resolution X-ray photoelectron spectroscopy (XPS) spectra of S 2p, N 1s, O 1s, C 1s, and Pb 4f were fitted using Avantage 5.997 software, as shown in Figure 10.

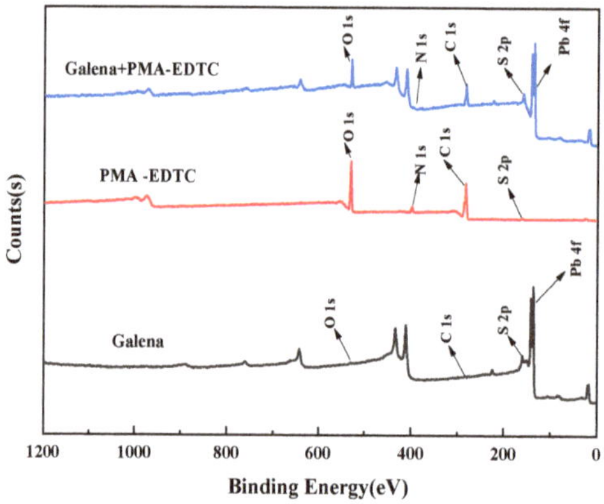

Figure 9. The wide survey scan XPS of untreated and treated galena with PMA-EDTC and PMA-EDTC complexes.

The high-resolution XPS spectrum of N 1s for PMA–EDTC and PMA–EDTC treated galena is presented in Figure 10. The N 1s spectrum of PMA–EDTC can be deconvoluted into two peaks with binding energies of 400.1 eV (–CONH$_2$ or –CONH–) and 401.7 eV (–NH–(C=S)–), consistent with reported values in the literature. After treatment with PMA–EDTC, two new peaks emerged at 399.5 eV and 400.4 eV on the N 1s spectrum of galena, corresponding to the (–CONH$_2$ or –CONH–) group and the (–NH–(C=S)–) group within PMA–EDTC. This confirms the adsorption of PMA–EDTC onto the galena surface, with shifts of 0.6 eV and 1.3 eV observed in these two peaks. These shifts validate the alteration of the chemical environment of N atoms within PMA–EDTC due to adsorption.

As depicted in the provided figure, in the high-resolution XPS spectra of S 2p for PMA–EDTC and galena, the characteristic S 2p3/2 peak and S 2p1/2 peak of pure galena appear at around 160.8 eV and 162.1 eV, respectively, with no other S 2p peaks observed.

This indicates the relative purity of the galena sample [37]. Additionally, the XPS spectrum of the S 2p peak for PMA–EDTC exhibits peaks at 162.5 eV, 163.6 eV, 164.1 eV, and 165.2 eV, attributed to the (C–S/C=S) group and the (–SH) group.

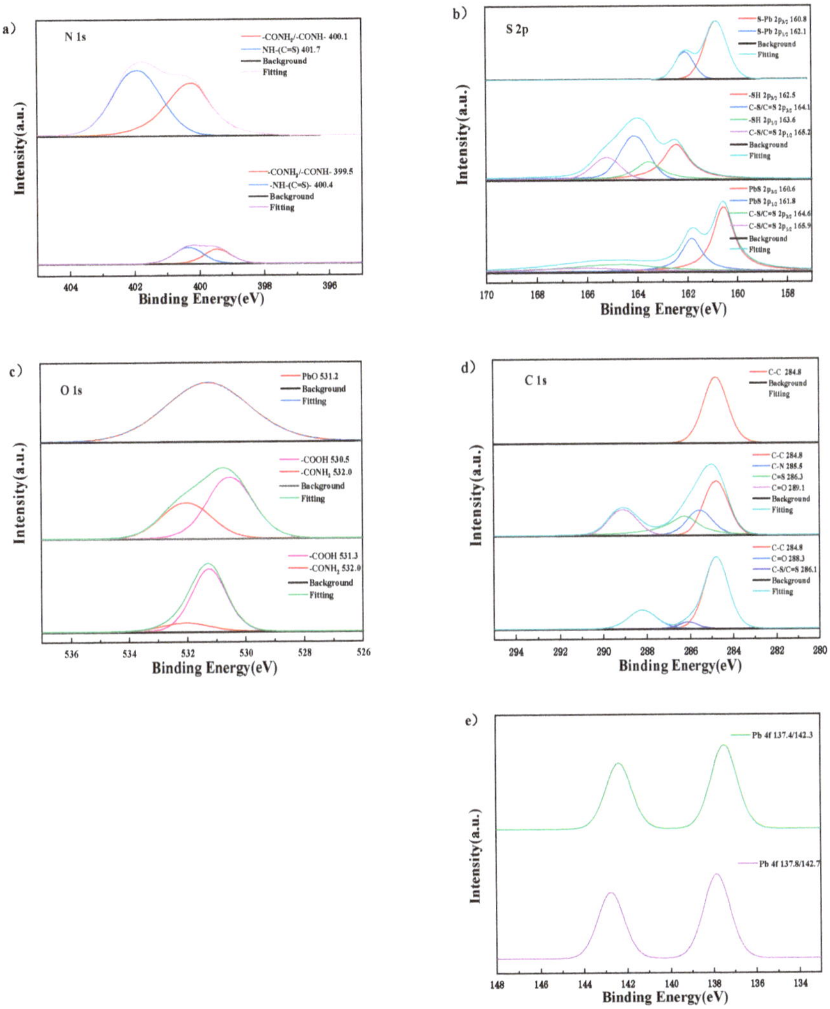

Figure 10. The fitted XPS of (**a**) N 1s, (**b**) S 2p, (**c**) O 1s, (**d**) C 1s, and (**e**) Pb 4f.

After treatment with PMA–EDTC, the surface of the galena displays four S 2p peaks at approximately 160.6 eV, 161.8 eV, 164.6 eV, and 165.9 eV. The first two S 2p peaks can be attributed to the S atoms of the original galena. The latter two peaks can be assigned to the (C–S/C=S) group within PMA–EDTC, based on comparison. Hence, it can be inferred that the di-thioamino acid root group (NH–C(=S)–S) in the structure of PMA–EDTC might react with the Pb sites on the galena surface to form complexes. This confirms that the inhibitor PMA–EDTC adsorbs onto the galena surface through the di-thioamino acid root group and forms Pb-S bonds with the surface Pb sites, resulting in the inhibition effect on the galena.

The high-resolution XPS spectrum of O 1s for galena reveals a single characteristic peak fitted at 531.2 eV, which can be attributed to potential surface oxidation. In contrast, the high-resolution O 1s XPS spectrum of PMA–EDTC displays two distinct peaks, positioned at 530.5 eV (–COOH) and 532.0 eV (–$CONH_2$), closely aligning with previously reported values [37]. Following the adsorption of PMA–EDTC onto the galena surface, the O 1s XPS high-resolution spectrum of galena shows two peaks appearing at approximately 531.3 eV and 532.1 eV, corresponding to the two peaks in the high-resolution O 1s spectrum of PMA–EDTC. The peak assigned to (–$CONH_2$) experiences minimal displacement, indicating relatively weak interactions of the (–$CONH_2$) group with the galena surface.

In the C 1s spectrum of PMA–EDTC, four distinct peaks are observed. The main peak at 284.8 eV corresponds to carbon contamination, while the other three peaks at 285.5 eV, 286.3 eV, and 289.1 eV correspond to PMA–EDTC's C–N, C=S/C–S, and C=O groups, respectively [38,39]. On the other hand, the C 1s spectrum of galena displays only one peak at 284.8 eV, also attributed to carbon contamination. In comparison to the original galena, the C 1s XPS high-resolution spectrum of galena after interaction with PMA–EDTC exhibits three peaks, at 286.1 eV and 288.3 eV, which correspond to PMA–EDTC's C=O and C=S/C–S groups, respectively. The shift of the C=S peak by 0.8 eV suggests the involvement of the (NH–C(=S)–S–) group in the reaction with Pb sites on the galena surface, leading to a change in the electron density of carbon atoms. This further confirms the adsorption of PMA–EDTC onto the galena surface.

In the graph of the high-resolution Pb 4f spectrum of pristine galena, two identified peaks located at 137.4 eV and 142.3 eV correspond to the characteristic Pb 4f peaks of galena, closely matching reported data [40,41]. Following treatment with PMA–EDTC, these two peaks underwent shifts of approximately 0.4 eV and 0.6 eV, respectively. The increase in the electron density of Pb might stem from alterations in the chemical state of Pb on the galena surface. This leads to further evidence that PMA–EDTC has interacted with Pb sites on the galena surface [42].

Combining the molecular structure of PMA–EDTC with the analyses of XPS and FI-IR, it can be inferred that the functional group in PMA–EDTC reacting with Pb sites on the galena surface is the NH–C(=S)–S– group. The adsorption model is illustrated as shown in Figure 11. As mentioned above, the PMA–EDTC structure contains a substantial number of hydrophilic groups, resulting in hydrophilic characteristics on the galena surface. This arrangement serves the purpose of inhibiting galena, effectively achieving the suppression of the galena flotation.

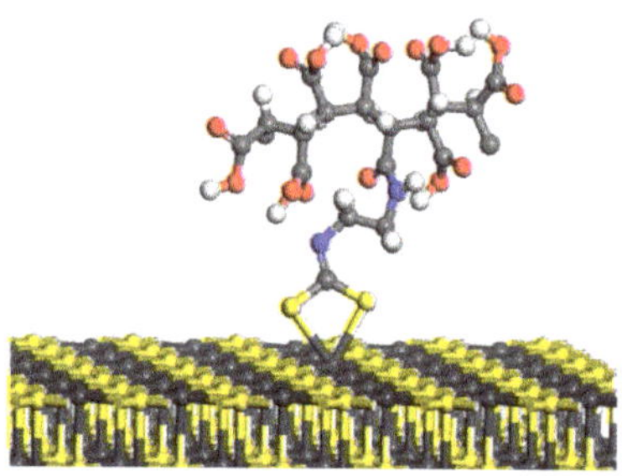

Figure 11. The recommended adsorption mode of PMA–EDTC on galena surface.

4. Conclusions

In this study, a novel macromolecular organic lead inhibitor (PMA–EDTC) was prepared by polymerizing maleic acid with NH–(C=S)–S– as a solid affinity group. The single mineral flotation test showed that PMA–EDTC had a good inhibitory effect on galena in a wide pH range of 2~12, but had little effect on the recovery of chalcopyrite. The results of the artificial ore mixing test showed that when the dosage of PMA–EDTC reaches 8 mg/L, the recovery rate of copper in the concentrate was 76.82%, while the recovery of lead was only 0.56%, and the copper-lead separation coefficient was 24.17, which further confirmed the effectiveness of PMA–EDTC as a lead inhibitor for copper-lead separation.

The FT-IR results indicated significant shifts and peak variations in the absorption peaks of the C=S–SH and $-NH_2$ functional groups of PMA–EDTC after interaction with Pb^{2+} ions, suggesting the occurrence of chemical reactions between PMA–EDTC and Pb^{2+} ions. Within the pH range of 2~12, treatment with PMA–EDTC resulted in a significant decrease in the Zeta potential of galena, indicating the adsorption of the reagent on the surface of galena and rendering it more hydrophilic. Contact angle measurements demonstrated that PMA–EDTC effectively altered the floatability of galena while having little impact on the floatability of chalcopyrite, achieving selective depression of lead and flotation of copper. XPS measurements further confirmed the proposed interaction model, showing that PMA–EDTC was chemically adsorbed on the lead atoms on the galena surface.

Author Contributions: Conceptualization, H.Z.; methodology, H.Z.,Y.Z. and C.S.; software, C.L. and T.L.; validation, Y.Z. and Z.Z.; formal analysis, T.L. and G.W.; investigation, H.Z. and G.W.; resources, X.Z. and Z.Z.; data curation, H.Z.; writing—original draft preparation, H.Z.; writing—review and editing, G.W and H.Z.; visualization, T.L. and G.W.; supervision, H.Z. and C.S.; project administration, C.L. and H.Z.; funding acquisition, Z.Z. and Y.Z. All authors have read and agreed to the published version of the manuscript.

Funding: This research was funded by the National Natural Science Foundation of China, grants: The interface regulation mechanism of efficient flotation separation of complicated nonferrous metal resources and the design of targeted reagent molecules (U20A20269).

Data Availability Statement: Data are contained within the article.

Conflicts of Interest: The authors declare no conflict of interest.

References

1. Cui, Y.; Jiao, F.; Qin, W.; Wang, C.; Li, X. Flotation separation of sphalerite from galena using eco-friendly and efficient depressant pullulan. *Sep. Purif. Technol.* **2022**, *295*, 121013. [CrossRef]
2. Zhang, M.; Li, J.; Li, P.; Liang, W.; Zheng, X.; Sun, K. Summarize on the lead and zinc ore resources of the world and China. *China Min. Mag.* **2016**, *25* (Suppl. S1), 41–45, 103.
3. Qin, W.; Wei, Q.; Jiao, F.; Yang, C.; Liu, R.; Wang, P.; Ke, L. Utilization of polysaccharides as depressants for the flotation separation of copper/lead concentrate. *Int. J. Min. Sci. Technol.* **2013**, *23*, 179–186. [CrossRef]
4. Cai, J.; Jia, X.; Ma, Y.; Ibrahim, A.M.; Su, C.; Yu, X.; Shen, P.; Liu, D. Effect of pre-oxidation on copper-lead bulk concentrate flotation separation with sodium polyacrylate as galena depressant. *Sep. Purif. Technol.* **2023**, *304*, 122276. [CrossRef]
5. Sehlotho, N.; Sindane, Z.; Bryson, M.; Lindvelt, L. Flowsheet development for selective Cu-Pb-Zn recovery at Rosh Pinah concentrator. *Miner. Eng.* **2018**, *122*, 10–16. [CrossRef]
6. Liu, R.; Lu, H.; Xu, Z.; Wang, C.; Sun, W.; Wu, M.; Dong, Y.; Bai, L. New insights into the reagent-removal mechanism of sodium sulfide in chalcopyrite and galena bulk flotation: A combined experimental and computational study. *J. Mater. Res. Technol.* **2020**, *9*, 5352–5363. [CrossRef]
7. Jiao, F.; Cui, Y.; Wang, D.; Hu, C. Research of the replacement of dichromate with depressants mixture in the separation of copper-lead sulfides by flotation. *Sep. Purif. Technol.* **2021**, *278*, 119330. [CrossRef]
8. Valdivieso, A.; Ledesma, L.A.; Cabrera, A.; Orozco Navarro, O.A. Carboxymethylcellulose (CMC) as PbS depressant in the processing of Pb Cu bulk concentrates. *Adsor. Float. Stud. Miner. Eng.* **2017**, *112*, 77–83. [CrossRef]
9. Huang, P.; Cao, M.; Liu, Q. Using chitosan as a selective depressant in the differential flotation of Cu–Pb sulfides. *Int. J. Miner. Process.* **2012**, *106*, 8–15. [CrossRef]
10. Pearse, M.J. An overview of the use of chemical reagents in mineral processing. *Miner. Eng.* **2005**, *18*, 139–149. [CrossRef]
11. Bulatovic, S.; Wyslouzil, D.M. Selection and evaluation of different depressants systems for flotation of complex sulphide ores. *Miner. Eng.* **1995**, *8*, 63–76. [CrossRef]
12. Zhang, X.; Lu, L.; Zeng, H.; Hu, Z.; Zhu, Y.; Han, L. A macromolecular depressant for galena and its flotation behavior in the separation from molybdenite. *Miner. Eng.* **2020**, *157*, 106576. [CrossRef]
13. Qin, W.; Wei, Q.; Jiao, F.; Li, N.; Wang, P.; Ke, L. Effect of sodium pyrophosphate on the flotation separation of chalcopyrite from galena. *Int. J. Min. Sci. Technol.* **2012**, *22*, 345–349. [CrossRef]
14. Chen, X.; Gu, G.; Chen, Z. Seaweed glue as a novel polymer depressant for the selective separation of chalcopyrite and galena. *Int. J. Miner. Metall. Mater.* **2019**, *26*, 1495–1503. [CrossRef]
15. Chen, J.; Lan, L.; Liao, X. Depression effect of pseudo glycolythiourea acid in flotation separation of copper–molybdenum. *Trans. Nonferrous Met. Soc. China* **2013**, *23*, 824–831. [CrossRef]
16. Yin, Z.; Sun, W.; Hu, Y.; Zhai, J.; Qingjun, G. Evaluation of the replacement of NaCN with depressant mixtures in the separation of copper–molybdenum sulphide ore by flotation. *Sep. Purif. Technol.* **2017**, *173*, 9–16. [CrossRef]

17. Liu, Y.; Liu, Q. Flotation separation of carbonate from sulfide minerals, II: Mechanisms of flotation depression of sulfide minerals by thioglycollic acid and citric acid. *Miner. Eng.* **2004**, *17*, 865–878. [CrossRef]
18. Li, M.; Wei, D.; Shen, Y.; Liu, W.; Gao, S.; Liang, G. Selective depression effect in flotation separation of copper–molybdenum sulfides using 2,3-disulfanylbutanedioic acid. *Trans. Nonferrous Met. Soc. China* **2015**, *25*, 3126–3132. [CrossRef]
19. Chen, J.; Feng, Q.; Lu, B. Research on a new organic depressant ASC for Separation chalcopyrite and galena. *Conserv. Util. Miner. Resour.* **2000**, *20*, 39–42. (In Chinese)
20. Piao, Z.; Wei, D.; Liu, Z.; Liu, W.; Gao, S.; Li, M. Selective depression of galena and chalcopyrite by O,O-bis(2,3-dihydroxypropyl)dithiophosphate. *Trans. Nonferrous Met. Soc. China* **2013**, *23*, 3063–3067. [CrossRef]
21. Pugh, R.J. Macromolecular organic depressants in sulphide flotation—A review, 1. Principles, types and applications. *Int. J. Miner. Process.* **1989**, *25*, 101–130. [CrossRef]
22. Kar, B.; Sahoo, H.; Rath, S.S.; Das, B. Investigations on different starches as depressants for iron ore flotation. *Miner. Eng.* **2013**, *49*, 1–6. [CrossRef]
23. Rath R, K.; Subramanian, S. Effect of guar gum on selective separation of sphalerite from galena by flotation. *Trans. Inst. Min. Metall. Sect. C—Miner. Process. Extr. Metall.* **1999**, *108*, C1–C7.
24. De Melo, B.A.G.; Motta, F.L.; Santana, M.H.A. Humic acids: Structural properties and multiple functionalities for novel technological developments. *Mater. Sci. Eng. C* **2016**, *62*, 967–974. [CrossRef]
25. Pearse, M.J. Historical use and future development of chemicals for solid–liquid separation in the mineral processing industry. *Miner. Eng.* **2003**, *16*, 103–108. [CrossRef]
26. Moudgil, B.M. Effect of polyacrylamide and polyethylene oxide polymers on coal flotation. *Colloids Surf.* **1983**, *8*, 225–228. [CrossRef]
27. Boulton, A.; Fornasiero, D.; Ralston, J. Selective depression of pyrite with polyacrylamide polymers. *Int. J. Miner. Process.* **2001**, *61*, 13–22. [CrossRef]
28. Zhang, J.F.; Hu, Y.H.; Wang, D.Z.; Han, L. Depressant effect of hydroxamic polyacrylamide on pyrite. *J. Cent. South Univ. Technol. (Engl. Ed.)* **2004**, *128*, 6–15.
29. Bulatovic, S.M. Use of organic polymers in the flotation of polymetallic ores: A review. *Miner. Eng.* **1999**, *12*, 341–354. [CrossRef]
30. Jiménez-Hernández, L.; Estévez-Hernández, O.; Hernández-Sánchez, M.; Díaz, J.; Sánchez, M.F.; Reguera, E. 3-mercaptopropionic acid surface modification of Cu-doped ZnO nanoparticles: Their properties and peroxidase conjugation. *Colloids Surf. A Physicochem. Eng. Asp.* **2016**, *489*, 351–359. [CrossRef]
31. Yin, Z.; Hu, Y.; Sun, W.; Zhang, C.; He, J.; Xu, Z.; Zou, J.; Guan, C.; Zhang, C.H.; Guan, Q.; et al. Adsorption mechanism of 4-amino-5-mercapto-1,2,4-triazole as flotation reagent on chalcopyrite. *Langmuir* **2018**, *34*, 4071–4083. [CrossRef]
32. Huang, P.; Wang, L.; Liu, Q. Depressant function of high molecular weight polyacrylamide in the xanthate flotation of chalcopyrite and galena. *Int. J. Miner. Process.* **2014**, *128*, 6–15. [CrossRef]
33. Fullston, D.; Fornasiero, D.; Ralston, J. Zeta potential study of the oxidation of copper sulfide minerals. *Colloids Surf. A Physicochem. Eng. Asp.* **1999**, *146*, 113–121. [CrossRef]
34. Vergouw, J.M.; Difeo, A.; Xu, Z.; Finch, J. An agglomeration study of sulphide minerals using zeta-potential and settling rate. Part 1: Pyrite and galena. *Miner. Eng.* **1998**, *11*, 159–169. [CrossRef]
35. Liu, R.; Qin, W.; Fen, J.; Wang, X.; Pei, B.; Yang, Y.; Lai, C. Flotation separation of chalcopyrite from galena by sodium humate and ammonium persulfate. *Trans. Nonferrous Met. Soc. China* **2016**, *26*, 265–271. [CrossRef]
36. Lu, L.; Xiong, W.; Zhu, Y.; Zhang, X.; Zheng, Y. Depression behaviors of N-thiourea-maleamic acid and its adsorption mechanism on galena in Mo-Pb flotation separation. *Int. J. Min. Sci. Technol.* **2022**, *32*, 181–189. [CrossRef]
37. Buckley, A.N.; Woods, R. An X-ray photoelectron spectroscopic study of the oxidation of chalcopyrite. *Aust. J. Chem.* **1984**, *37*, 2403–2413. [CrossRef]
38. Yin, Z.; Sun, W.; Hu, Y.; Zhang, C.; Guan, Q.; Liu, R.; Chen, P.; Tian, M. Utilization of acetic acid-[(hydrazinylthioxomethyl) thio]-sodium as a novel selective depressant for chalcopyrite in the flotation separation of molybdenite. *Sep. Purif. Technol.* **2017**, *179*, 248–256. [CrossRef]
39. Li, W.; Li, Y. Improved understanding of chalcopyrite flotation in seawater using sodium hexametaphosphate. *Miner. Eng.* **2019**, *134*, 269–274. [CrossRef]
40. Ikumapayi, F.; Makitalo, M.; Johansson, B.; Rao, K.H. Recycling of process water in sulphide flotation: Effect of calcium and sulphate ions on flotation of galena. *Miner. Eng.* **2012**, *39*, 77–88. [CrossRef]
41. Yu, J.; Liu, R.; Wang, L.; Sun, W.; Peng, H.; Hu, Y. Selective depression mechanism of ferric chromium lignin sulfonate for chalcopyrite–galena flotation separation. *Int. J. Miner. Metall. Mater.* **2018**, *25*, 489–497. [CrossRef]
42. Wang, Y.; Xiong, W.; Zhang, X.; Lu, L.; Zhu, Y. A new synthetic polymer depressant PADEMA for Cu-Pb separation and its interfacial adsorption mechanism on galena surface. *Appl. Surf. Sci.* **2021**, *569*, 151062. [CrossRef]

minerals MDPI

Article

Application and Mechanism of Mixed Anionic/Cationic Collectors on Reverse Flotation of Hematite

Juanjuan Wang, Dongfang Lu * and Xue Han

School of Minerals Processing & Bioengineering, Central South University, Changsha 410083, China; 215611054@csu.edu.cn (J.W.); 17554215224@163.com (X.H.)
* Correspondence: ludongfang@csu.edu.cn

Abstract: In order to enhance the reverse flotation effect of hematite, dodecyl trimethyl ammonium chloride (DTAC) and a mixed anionic/cationic collector of DTAC and tall oil were selected for flotation studies on quartz and hematite. Surface tension tests, FT-IR, XPS analysis, and molecular dynamics (MD) simulations were also conducted in order to investigate the interaction mechanism of the mixed collector on the quartz surface. The results revealed that, at a natural pH (approximately 7.0) and 298 K, and with a dosage of 20 mg/L for the mixed collector (m_{DTAC}:$m_{tall\ oil}$ = 3:1) and 4 mg/L for causticized starch, the recovery rates of quartz and hematite were 94.67% and 8.69%, respectively. Compared to the use of a single DTAC, the mixed DTAC/tail oil collector enhanced the flotation effect under constant variables. Additionally, a comparison of temperature and surface tension for monomineral flotation and artificially mixed ore flotation tests showed that the mixed collector was less influenced by temperature, compared to the single DTAC, and exhibited good resistance and more effective separation of quartz and hematite at lower temperatures. The analysis of FT-IR, XPS, and MD simulations revealed that the single DTAC collector primarily adsorbed onto quartz through physical adsorption and hydrogen bonding, while the combination of tall oil and DTAC enhanced the adsorption of the collector on the quartz surface.

Keywords: quartz; hematite; mixed collector; reverse flotation; molecular dynamics simulation

1. Introduction

Steel is widely used in the daily lives of human beings, as transportation, housing construction, and home appliance manufacturing, among other industries, all rely heavily on the supply of the steel industry. The rapid expansion of the steel industry is inevitably leading to a surge in the consumption of iron ore resources, with nearly 98% of the world's iron ore being utilized for steel production [1]. Currently, the depletion of easily accessible high-grade iron ore reserves has prompted the exploration for alternatives in deeper, lower- to intermediate-grade ores, which are complex and challenging to process in order to satisfy increasing demands [2–4]. Research on the technology of "iron extraction and silicon reduction" has pointed out the development direction for improving the separation index of iron ore in China and reducing production costs. In recent years, through the concerted efforts of mineral processing professionals, significant progress has been made in the beneficiation technology of low-grade hematite. The key technologies of "iron extraction and silicon reduction" include the applications of reverse flotation technology, new magnetic separation equipment, and fine sieve technology, as well as the employment of flotation columns. Among these technologies, the application of reverse flotation technology is the research focus in the field of iron ore flotation.

Reverse flotation technologies include anionic reverse flotation and cationic reverse flotation [5,6]. Anionic reverse flotation was initially developed in China, where anionic collectors selectively adsorb onto the surface of iron ore through specific chemical reactions, enabling the separation of ore from gangue [7]. Oleic acid, among other fatty acid anionic surfactants, is often used. Studies have shown that, when Ca^{2+} is used as an activator, NaOL shows good harvesting and selectivity to quartz under pH conditions of about 11.5. However, the chemical system of anionic reverse flotation technology is more complex; the slurry needs to be heated to more than 308 K under industrial conditions, the production cost is high, and a higher pH corrodes the equipment [8,9]. Advancements in research reveal that cationic collectors excel in treating certain iron ores, leading reverse flotation technology to focus on optimizing processes and procedures, enhancing concentrated iron quality, increasing recovery rates, and reducing harmful impurities, such as phosphorus and sulfur, thus surpassing other beneficiation methods [10]. In iron ore cationic reverse flotation, the commonly used cationic reverse flotation collector is mostly dodecyl amine, which is different from that used in anionic reverse flotation technology; when using cationic reverse flotation technology to separate quartz and iron minerals, no additional activator is needed to show good recovery of quartz under weak alkaline conditions, but selectivity is poor and the cost is high. In addition, due to the strong frothing ability of dodecyl amine, the problem of defrothing also needs to be solved urgently [11–13]. With the continuous improvement in energy conservation and environmental protection requirements, the problems of both high preparation temperatures and high-use temperatures for iron ore flotation reagents have become increasingly prominent, seriously restricting the development of domestic iron mines. In recent years, through a large number of studies on iron ore reverse flotation technology, it has been found that research on flotation reagents, especially high-efficiency collectors, is still key to improving flotation efficiency. Some novel collecting agents exhibit excellent selectivity, but are challenging and costly to synthesize, while the use of a mixed collector not only minimizes reagent consumption and production costs but also addresses the shortcomings associated with using a single reagent, thereby further boosting flotation efficiency [14–16]. For example, Yang et al. [17] found that the mixed collector of sodium oleate–benzohydroxamic acid (NaOL–BHA), at a molar ratio of 4:1, exhibited excellent performance in the flotation separation of ilmenite from titanaugite. Tian et al. [18] demonstrated that, in the processing of lithium pegmatite ore, utilizing a combination of mixed anionic/cationic collectors, SX (laurylamine and oxidized paraffin wax soap) and MOD (laurylamine and naphthenic acid soap), led to the enhanced grade and recovery of Li2O in spodumene concentrates, achieving levels of 6.02% and 87.34%, respectively.

In this study, reverse flotation tests were carried out on hematite and quartz with dodecyl trimethyl ammonium chloride (DTAC) and an anionic and cationic mixed collector mainly based on DTAC; the flotation behaviors of quartz and hematite under different pH, dosages, depressors, and temperatures, were compared and investigated, and then a surface tension test, Fourier transform infrared (FTIR) spectroscopy, X-ray photoelectron spectroscopy (XPS), and molecular dynamics simulations were conducted in order to compare and analyze the mixed collector's mechanism of action, which provided a theoretical basis for the reverse flotation separation of quartz and hematite at room, or low temperature.

2. Materials and Methods

2.1. Mineral Samples and Reagents

The samples of hematite and quartz used in the experiments were high-purity lump ores, purchased from the market, crushed, hand-separated, and ground with an iron hammer. Subsequently, they were screened and graded using 75 μm and 38 μm Taylor standard sieves. The ore samples with particle sizes of −75 + 38 μm were utilized for flotation tests, while ore samples with a particle size of −38 μm were employed for analysis and detection. The material compositions and surface properties of the hematite and quartz monomineral samples of the two particles' sizes were identical. The X-ray fluorescence (XRF) analysis and X-ray diffraction (XRD) analysis results of quartz and hematite are shown in Table 1, Figure 1, and Figure 2. The results indicated that the purity of the test ore samples exceeded 95%, thus meeting the requirements of the monomineral test.

Table 1. Chemical compositions of quartz and hematite (%).

Minerals	TFe	FeO	Al_2O_3	SiO_2	K_2O	Na_2O	Fe_2O_3	CaO	MgO
hematite	68.43	0.75	0.59	3.54	0.05	-	97.84	0.03	0.13
quartz	-	-	1.21	98.36	0.18	0.036	0.02	0.04	0.03

TFe is the total amount of iron, determined through the chemical analysis of ore samples.

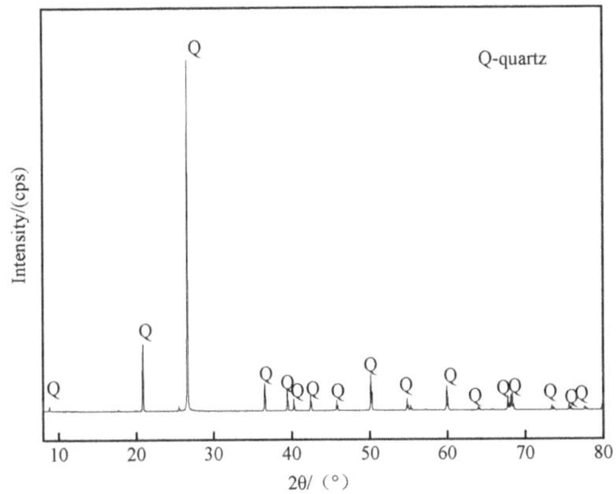

Figure 1. XRD patterns of quartz.

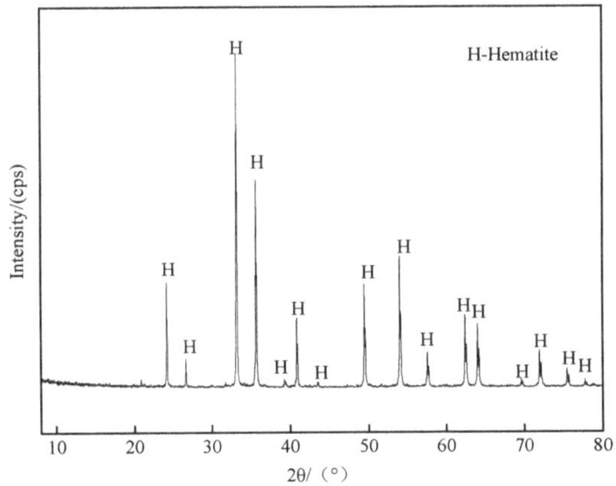

Figure 2. XRD patterns of hematite.

The DTAC and pH adjusters (NaOH and HCl) used in the test were all analytically pure, the causticized corn starch (corn starch and sodium hydroxide, mixed with sodium hydroxide according to a certain mass ratio, and causticized in a water bath at about 363 K for 30 min) was made with industrial-grade products, and the mixed collector of tall oil and DTAC (hereinafter referred to as "the mixed collector") was laboratory-prepared at a mass ratio of 1:3 and subsequently utilized. Deionized water (18.2 MΩ·cm) was used for both the flotation tests and the investigation of the agent's mechanism of action.

2.2. Micro-Flotation Tests

The flotation experiment was carried out using an XFG hanging tank flotation machine (Figure 3) from the Jilin Exploration Machinery Plant in Changchun, China, with a spindle speed of 1992 r/min. Each time, 3.0 g of the monomineral quartz or hematite was weighed and placed in a 50 mL flotation tank, and an appropriate amount of DI water was added and mixed for 2 min. Thereafter, the pH was adjusted with HCl or NaOH for 2 min. The depressor was added and allowed to act for 3 min (this step was only required for depressor tests, not for other condition tests). Collectors were then added and allowed to act for 2 min. Subsequently, the froth was scraped for 5 min. After the flotation process was complete, the froth product and the product in the tank were filtered, dried, and weighed, and the recovery value was calculated. The flotation test process is shown in Figure 4.

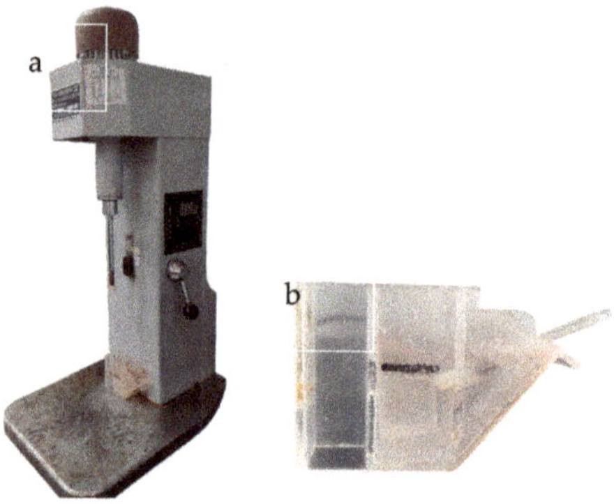

Figure 3. (**a**) Diagram of the XFG flotation machine; (**b**) flotation cell.

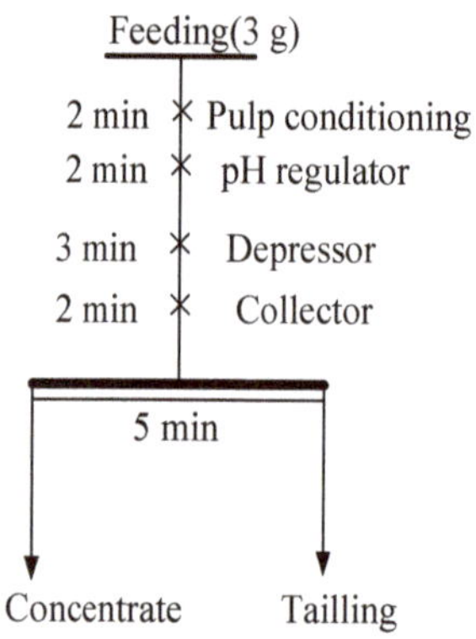

Figure 4. Flotation flowsheet of micro-flotation tests and artificially mixed mineral flotation tests.

2.3. *Flotation Experiments of Artificially Mixed Minerals*

In artificially mixed mineral flotation experiments, the weight ratio of hematite to quartz in the raw ore was 2.66: 0.34. The total iron (TFe) grade of the artificially mixed mineral was approximately 60.67%. The slurry mixing process mirrored that of the monomineral flotation, as depicted in Figure 4. The concentrate and tailings were filtered and dried post-flotation, the TFe grade was analyzed, and the recovery value was calculated using Equations (1) and (2):

$$\gamma = \frac{M_1}{M_1 + M_2} \times 100\% \tag{1}$$

where, "γ" is the flotation productivity, /%; "M_1" is the concentrate mass/g; "M_2" is the tailing mass/g. Equation (2) is as follows:

$$\varepsilon = \gamma \frac{\beta_1}{\beta_0} \times 100\% \tag{2}$$

where, "ε" is the flotation recovery, /%; "β_1" is the concentrate grade, /%; "β_2" is the artificially mixed mineral grade, /%.

2.4. Surface Tension Test

The surface tension of the solution was determined using the BZY-2 surface tension meter, employing the platinum plate method. When the platinum plate was immersed in the liquid to be measured, it was affected by the pull-down effect of the surface tension of the solution. Once equilibrium was achieved, the instrument provided the surface tension value corresponding to the immersion depth of the platinum plate. Each measurement was replicated five times, and the average was calculated.

2.5. Fourier Transform Infrared Spectroscopy Measurements

FT-IR of the samples was measured with a Nicolet 380 FT-IR analyzer (Thermo Fisher, Waltham, MA, USA) using the pellet method. The mineral sample, with a particle size of −38 μm, was placed in a flotation cell and prepared according to the flotation process shown in Figure 3. Following three washes with deionized water and filtration, the sample was dried in a vacuum drying oven before being submitted for FT-IR analysis. The spectral wavenumber range was set from 4000 to 400 cm^{-1}, and the solids spectrum was measured using the disk method, with KBr as the matrix [19,20].

2.6. X-ray Photoelectron Spectroscopy (XPS) Measurements

X-ray photoelectron spectroscopy (XPS) analysis samples were prepared following the same procedure as for FT-IR analyses. Detection was performed utilizing a Thermo Scientific K-Alpha X-ray photoelectron spectrometer, fitted with Advantage software, and calibrated based on a 284.80 eV C1s spectrum.

2.7. Molecular Dynamics Simulation

Molecular simulations were primarily conducted using Material Studio 2019 software, with a focus on the Dmol3 module. The collector molecule was chosen to adsorb on the (101) surface of quartz, referencing the existing literature [21–26]. In the Forcite module, an extended optimized surface unit cell of 2 × 3 × 1 was established, based on the COMPASS force field, with a vacuum layer thickness of 30 Å. The optimized crystal structure of quartz and the molecular structure of collectors are shown in Figures 5–7.

The optimized agent molecules were placed on the surface of the quartz mineral, and geometric optimization was carried out again in order to determine the adsorption model of the collector. The adsorption energy and geometric configuration were used to predict and analyze the interaction mechanism between the collector and the quartz mineral surface. The interaction energy is calculated as follows:

$$E_{ads} = E_{surf/dep} - E_{surf} - E_{dep} \tag{3}$$

where E_{ads} is the adsorption energy of the collector and quartz mineral surface; $E_{surf/dep}$ is the energy of the optimized system; E_{surf} is the total energy of quartz before and after adsorption by the collector; E_{dep} is the total energy of free collector molecules. A negative value of the interaction energy indicates the force of attraction between the mineral and the reagent, while a positive value indicates repulsion between the mineral and the reagent. The lower the E_{ads} value, the stronger the adsorption effect of the agent molecules on the surface of quartz minerals, while a positive E_{ads} value indicates that adsorption between the agent molecules and the surface of quartz minerals does not occur [27].

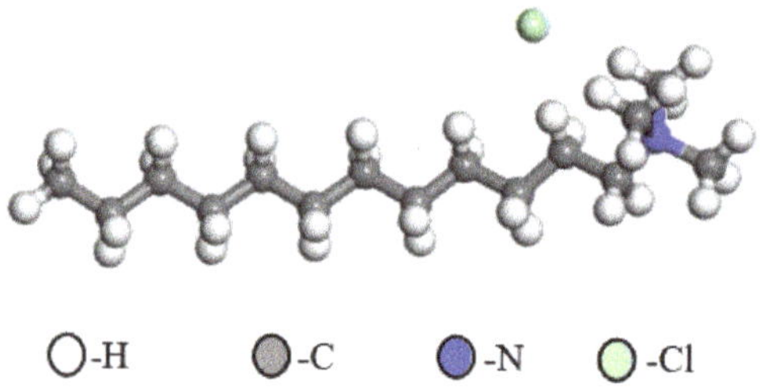

Figure 5. Structure of cationic DTAC.

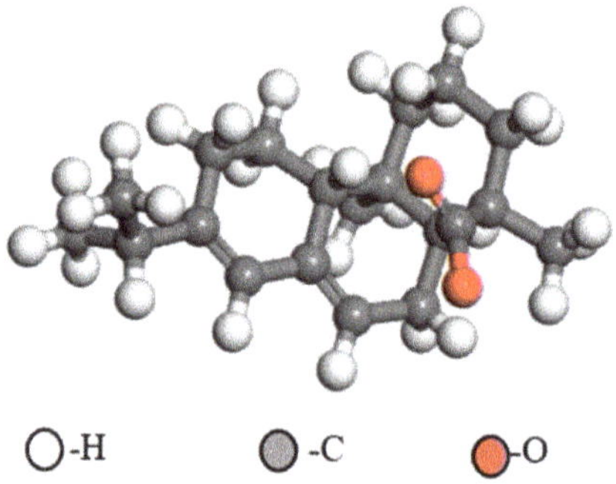

Figure 6. Structure of anionic tall oil.

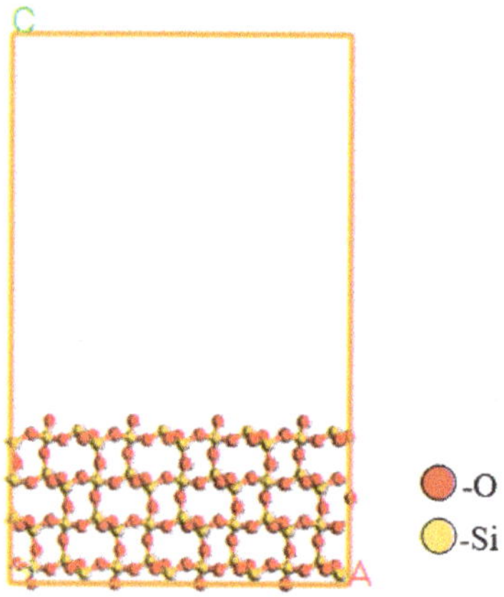

Figure 7. Crystal structure of quartz after optimization.

3. Results

3.1. Micro-Flotation Tests

The relationships between the recovery rates of quartz and hematite and the pH of the slurry are shown in Figure 8, where it can be seen that the recovery rates of quartz and hematite initially increased, and then decreased with a rise in pH under the same collector dosage conditions. The pH significantly influenced the recovery of quartz, with both collectors demonstrating good capture performance for quartz at pH levels between 6 and 10. The mixed collector exhibited relatively better recovery performance, achieving a quartz recovery value of up to 97.84%. In the pH range tested, both collectors displayed poor recovery performance for hematite. When compared to using a single DTAC, the mixed collector resulted in lower hematite recovery values, approximately 15%. Notably, in neutral and weakly alkaline slurry environments, both DTAC and the mixed collector effectively separated both hematite and quartz. In the comparative analysis, it was found that, under the same dosage conditions, the mixed collector exhibited higher selectivity, had a broader pH adaptability range, and was more conducive to the separation of quartz and hematite. Subsequent experiments were carried out without adding a pH adjuster, maintaining a natural pH (approximately 7.0) environment.

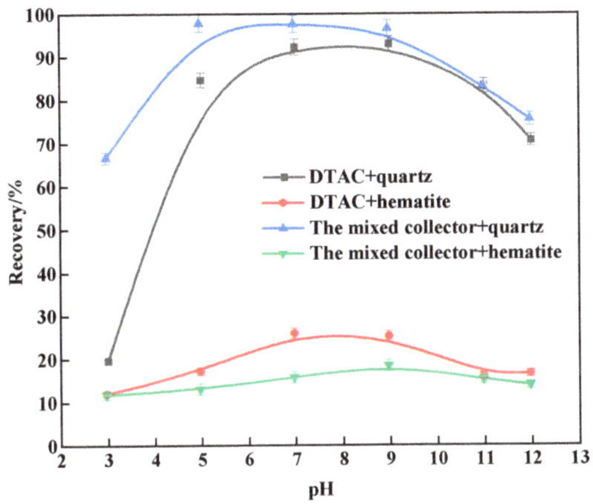

Figure 8. Effect of pH on flotation recovery rates of quartz and hematite using different collectors.

The relationship between the dosages of DTAC and the mixed collector on the recovery rate of quartz under the natural pH (approximately 7.0) condition of the slurry is shown in Figure 9. It was observed that the recovery rates of quartz and hematite increased as the dosage was raised within the measured range, and when the dosage of the agent was ≥20 mg/L, the recovery value of quartz in both collector systems exceeded 90%, with the recovery value using the mixed collector significantly surpassing that with the single DTAC. Compared with the mixed collector, DTAC had a greater impact on the recovery value of hematite with an increase in the collector dosage. The selectivity of the mixed collector was noted to be stronger than that of DTAC, with relatively good selectivity at 20 mg/L.

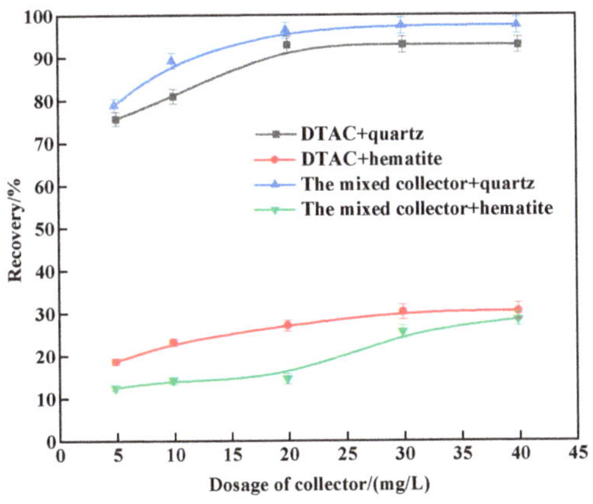

Figure 9. Effect of collector dosage on flotation recovery value of quartz and hematite under natural pH condition.

Figure 10 shows the effect of the causticized corn starch dosage on the recovery rates of quartz and hematite under the conditions of natural pH (approximately 7.0), fixed DTAC, and a mixed collector dosage of 20 mg/L, and it was observed that, as the dosage of causticized corn starch increased, the recovery rates of both quartz and hematite exhibited decreasing trends. The influence of the causticized corn starch dosage on the floatability of minerals remained consistent for both agents. The impact of causticized corn starch on the recovery of quartz was not pronounced. At a causticized corn starch dosage of 4 mg/L, the recovery rate of hematite was approximately 10%, and the recovery value showed minimal variation with further increases in starch dosage. As the causticized corn starch dosage increased, from 0 to 4 mg/L, the recovery value of hematite decreased, by approximately eight percentage points, to 8.69%. Subsequent increases in the amount of causticized corn starch did not significantly alter the recovery value of hematite. In the comparative analysis, it was found that, under other identical conditions, the addition of depressors had a greater impact on the flotation recovery of hematite. Furthermore, this study revealed that the addition of depressors was more beneficial for enhancing the selectivity of the mixed collector.

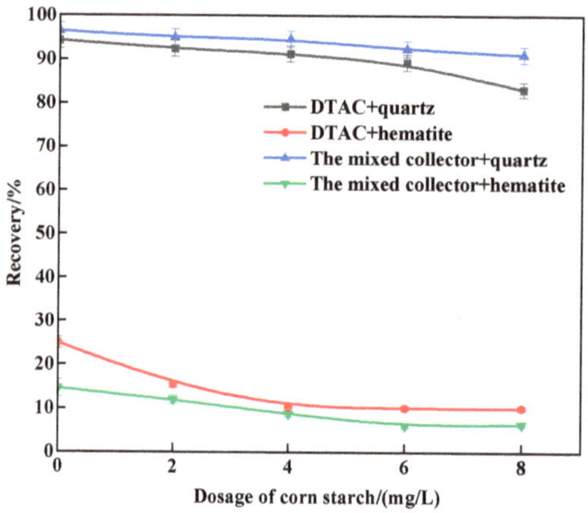

Figure 10. Effect of corn starch dosage on flotation recovery values of quartz and hematite under natural pH condition.

Figure 11 shows the effects of temperature on the recovery values of quartz and hematite under the natural pH condition (approximately 7.0) of the slurry, fixed DTAC, and a mixed collector dosage of 20 mg/L. As shown in Figure 11a, as the temperature decreased, the impact of the mixed collector on the recovery value of quartz was minimal, remaining around 90%. In contrast, the single DTAC collector exhibited a more pronounced effect on the recovery value of quartz with temperature variations, dropping below 80% when the temperature was below 283 K. In Figure 11b, the recovery value of hematite showed insignificant changes with temperature variations under the mixed collector system and slight decreases with decreasing temperature under the single DTAC collector system. Combined with Figure 11a,b, it can be inferred that the single DTAC was more sensitive to slurry temperature, and the mixed collector was less influenced by temperature, demonstrating good adaptability to low-temperature conditions.

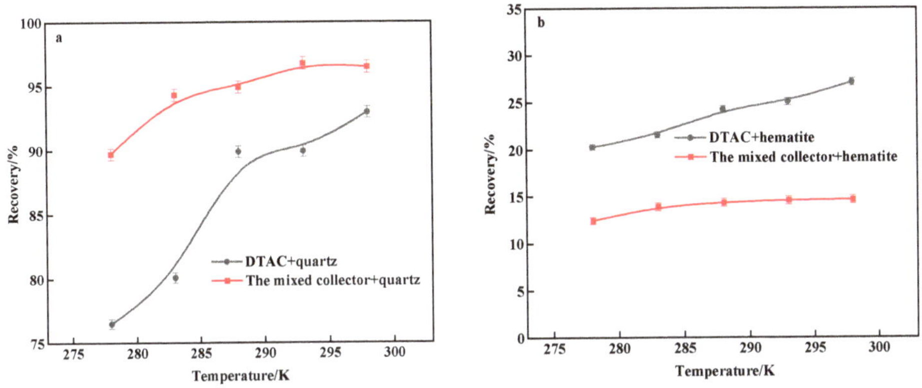

Figure 11. Effect of temperature on flotation recovery values of (**a**) quartz and (**b**) hematite under natural pH conditions.

3.2. Flotation Test of Artificially Mixed Minerals

The monomineral flotation test indicated that, in comparison to the single DTAC, the mixed collector showed better low-temperature flotation performance. To further validate this conclusion, an artificially mixed mineral flotation test was conducted at various temperatures. The test results are shown in Table 2.

Table 2. Flotation index of artificially mixed ore (the mixed collector: 20 mg/L; DTAC: 20 mg/L; corn starch: 4 mg/L).

Collector	Temperature, K	Product	Yield (%)	Grade, TFe (%)	Recovery (%)
The mixed collector	298	Concentrate	88.72	66.49	97.23
		Tailing	11.28	14.90	2.77
		Feed	100.00	60.67	100.00
	288	Concentrate	88.98	66.27	97.19
		Tailing	11.02	15.47	2.81
		Feed	100.00	60.67	100.00
DTAC	298	Concentrate	90.00	65.21	96.73
		Tailing	10.00	19.84	3.27
		Feed	100.00	60.67	100.00
	288	Concentrate	91.12	63.04	94.68
		Tailing	8.88	36.35	5.32
		Feed	100.00	60.67	100.00

As shown in Table 2, the selectivity and harvesting performance of the mixed collector were found to be stronger than those of the single DTAC when a depressor was added and other conditions remained unchanged, which corresponds to the monomineral flotation test. When the mixed collector was chosen for the reverse flotation of the artificial mixture, the flotation index of the concentrate showed minimal variation at 288 K and 298 K, with a flotation recovery value of approximately 97.20%. In contrast, when the single DTAC was used as the collector, the agent selection and capture performance decreased at 288 K, leading to a decrease in the TFe grade, by around two percentage points. These results confirmed that the mixed collector performed better than DTAC for flotation at lower temperatures.

3.3. The Effect of Temperature on the Surface Tension of the Collector

The surface tension values of the DTAC solution and the mixed collector at 298 K are shown in Figure 12. In comparison with the single DTAC system (12.94 mN/m,

1.00 g/L), the minimum surface tension and CMC value of the mixed collector system (13.64 mN/m, 0.015 g/L) were slightly lower, indicating higher surface activity and better foaming properties, but with a larger CMC value and weaker flotation capture capacity. The former exhibited lower surface tension and CMC values, and better hydrophobic performance of the agent, suggesting that the combined system had a better capture capacity, compared to a single agent, which is consistent with the flotation results. Therefore, concentrations of 0.015 g/L for the mixed collector solution and 1.00 g/L for the DTAC solution were determined as the optimal concentrations for further experimentation [7]. As shown in Figure 13, the surface tension decreased from 21.24 mN/m to 16.64 mN/m as the temperature of the 0.015 g/L mixed collector solution increased from 278 K to 298 K; meanwhile, the surface tension decreased from 23.07 mN/m to 11.94 mN/m as the temperature of the 1.00 g/L DTAC solution increased from 278 K to 298 K, indicating that the surface tension of the DTAC solution underwent a more significant change than that of the mixed collector with increasing temperature. This suggests that the mixed collector exhibited stronger temperature adaptability and was more conducive to low-temperature reverse flotation.

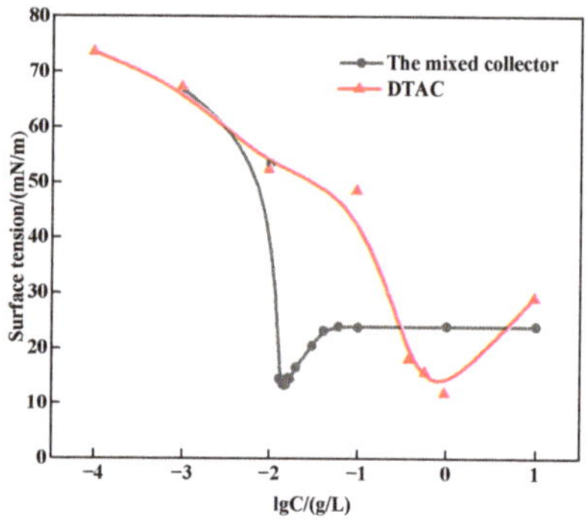

Figure 12. Relationship between surface tension and logarithm of concentration of collector solution.

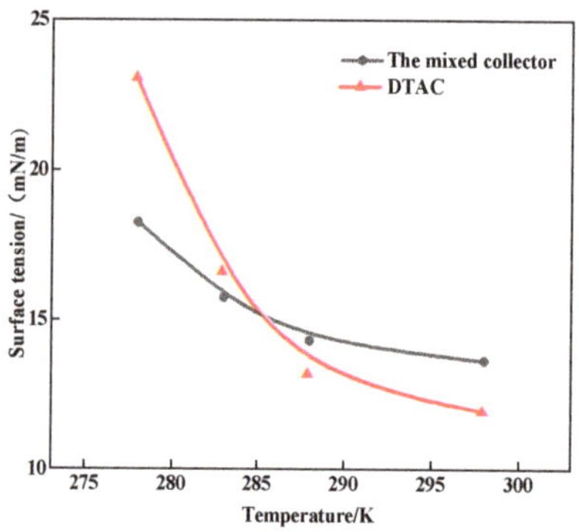

Figure 13. Effect of temperature on surface tension of collector.

3.4. FT-IR Analyses

To further investigate the adsorption mode of the mixed collector on the quartz surface, the infrared spectra of quartz, DTAC, and the agent before and after interaction were analyzed. The results are shown in Figure 14. From the figure, it can be observed that the characteristic wavenumbers of quartz were 1880.26, 1081.48, 796.32, 692.22, and 459.69 cm^{-1}. Among them, 1880.26 cm^{-1} corresponded to the tensile peak of the silicon/oxygen tetrahedron, 1081.48 cm^{-1} corresponded to the asymmetrical stretching vibration absorption peak of Si-O, and 796.32 and 692.22 cm^{-1} corresponded to the symmetrical stretching vibration absorption peak of Si-O-Si.

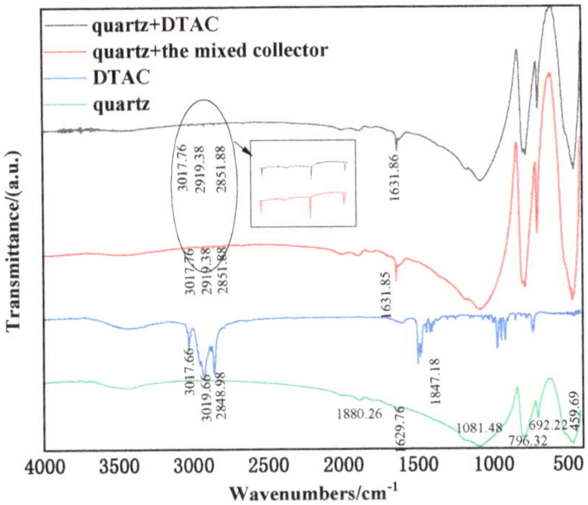

Figure 14. FT-IR spectra of quartz before and after treatment with reagents.

After the action of DTAC, characteristic peaks of DTAC were observed at 3018.76, 2919.38, 2848.98, and 1631.86 cm^{-1}. Specifically, the peak at 3018.76 cm^{-1} corresponded to the symmetrical stretching vibration absorption peaks of unsaturated -CH, while the peaks at 2919.38 cm^{-1} and 2848.98 cm^{-1} were attributed to the asymmetric and symmetrical stretching vibration absorption peaks of -CH_2 and -CH_2 from DTAC, respectively, indicating the occurrence of DTAC adsorption on the quartz surface. Additionally, the bending vibration absorption peak of -OH in ≡SiOH on the quartz surface at 1631.86 cm^{-1} became more prominent, possibly due to the formation of strong hydrogen bonds between the -NH- groups in the reagent structure and the O atoms or -OH groups in the ≡SiOH groups on the quartz surface. When quartz was treated with the mixed collector, the same absorption peak appeared as that observed after DTAC action, indicating that physical adsorption and hydrogen bonding action also took place on the surface of the quartz with the mixed collector. However, the intensity of the peaks was more pronounced, and the peak shape around 1081 cm^{-1} was wider compared to that of quartz after DTAC action, suggesting stronger adsorption of the mixed collector on the surface of the quartz mineral.

3.5. XPS Analysis

XPS tests are utilized to accurately determine the chemical environment in which an element is exposed and its content. To further investigate the adsorption mechanism of the mixed collector on the quartz surface, XPS tests were conducted on the quartz samples both before and after adsorption. The displacement of the binding energy of the elements present in the mixed collector and the alterations in the relative concentration of each atom were analyzed in detail.

Table 3 reveals a decrease in Si content, an increase in C and O content, and the presence of N in both quartz and the collector, indicating the occurrence of agent adsorption on the mineral surface. Among them, the N elements on the quartz surface following the action of the mixed collector and DTAC were 0.66 and 0.26, respectively (exceeding the instrument error). The N concentration on the quartz surface treated with the mixed collector was significantly higher than that on the quartz surface treated with DTAC, demonstrating the superior adsorption capacity of the mixed collector on the quartz surface compared to that of DTAC. This finding supported the enhanced capture performance of the mixed collector, aligning with the flotation outcomes of the monomineral test.

Table 3. The different atomic concentrations of quartz before and after the action of chemicals.

Sample	Atomic Concentration (%)			
	C1s	O1s	Si2p	N1s
Quartz + deionized water	13.39	56.45	29.62	-
Quartz + the mixed collector	14.46	56.87	28.01	0.66
Quartz + DTAC	14.08	56.45	28.91	0.26

To further investigate the adsorption mechanism of the mixed collector on the quartz surface, a narrow spectrum of N1 elements was depicted, as illustrated in Figure 15. In Figure 15a, no N1 peaks were observed in the narrow spectrum of pure quartz minerals. Upon the introduction of a single DTAC collector, new peaks emerged. One peak appeared around 398.07 eV, with a relative content of 0.18%, associated with the neutral dodecyl molecule (RH_2) interacting with the quartz mineral surface, primarily through hydrogen bonding; another peak, at approximately 400.03 eV, with a relative content of 0.08%, was linked to the protonated dodecyl group (RNH_3^+) interacting with the quartz mineral surface, mainly through physical adsorption [28]. A comparison of Figure 15b,c revealed that, upon the simultaneous addition of tall oil, no new peaks surfaced. The binding energies of the two peaks were 398.05 eV and 401.05 eV, with relative contents of 0.46% and 0.20%, respectively, remaining largely unchanged compared to the addition of a single DTAC collector (within the instrument error range). This suggests that the addition of tall oil did not involve chemical adsorption with the quartz surface. Nonetheless, the total relative N content in quartz treated with the mixed collector, as shown in Table 3, increased to 0.66%, nearly three times that of N after treatment with a single DTAC, indicating a greater adsorption capacity of the mixed collector on the quartz mineral surface. This finding aligned with the results from infrared spectroscopy.

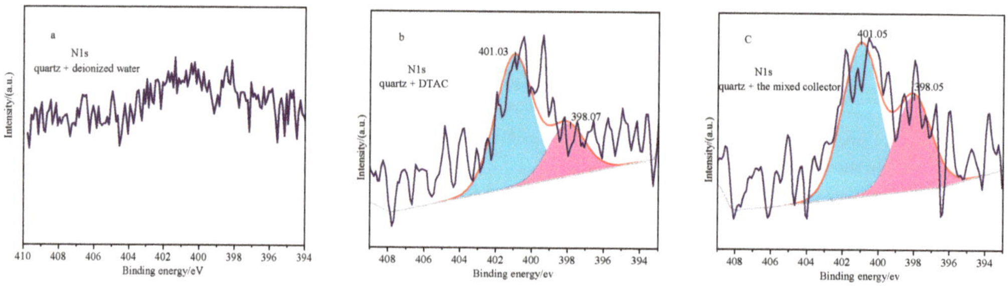

Figure 15. N1 spectra on the surface of a quartz sample treated under (**a**) deionized water, (**b**) DTAC, and (**c**) the mixed collector.

3.6. DFT Analyses

The adsorption behaviors of DTAC and the mixed collector on the quartz surface were compared and analyzed using MS. Four reagent molecules were introduced to the quartz (101) surface and optimized geometrically with the Forcite module. The van der Waals force and electrostatic energy were calculated utilizing the atomic and the Ewald summation methods. The most stable configurations, as shown in Figure 16a,b, were selected based on the minimum adsorption energy. It is important to note that, since this reverse flotation process is primarily conducted in a natural solution (approximately 7.0), the agent model chosen for the MD simulation consisted of dissociated anionic surfactant ions and undissociated DTAC molecules.

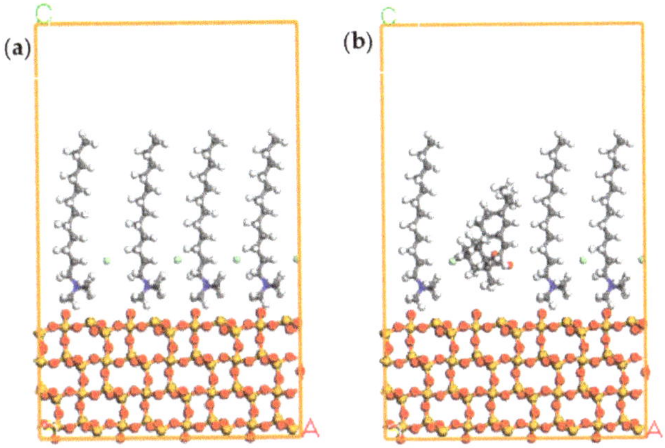

Figure 16. Initial adsorption models of (**a**) DTAC and (**b**) the mixed collector onto the (101) quartz surface.

From Figure 17a,b, it can be seen that the hydrophilic head group of DTAC exhibited uniform and stable adsorption on the quartz surface, with the hydrophobic carbon chains clustering accordingly. The aggregation phenomenon in Figure 17b was more pronounced. When combined with the results from Table 4, it became apparent that both collectors adsorbed onto the quartz mineral surface. Specifically, the adsorption energy of the mixed collector on the quartz surface was −52.16 kcal/mol higher than that of DTAC (−35.10 kcal/mol) on the mineral surface, indicating stronger adsorption of the mixed collector on quartz. This observation suggested that the total adsorption energy of the collector could be significantly enhanced through co-adsorption, a finding consistent with the results of the FT-IR and XPS analyses.

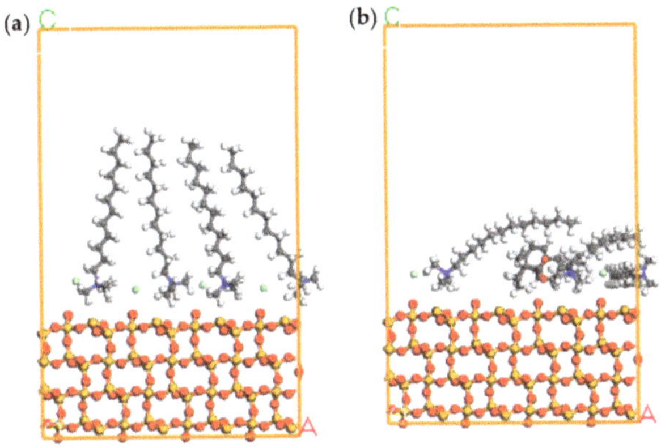

Figure 17. MD simulations of (**a**) DTAC and (**b**) the mixed collector onto the (101) quartz surface.

Table 4. The interaction energies of collectors onto the (101) quartz surface.

Collector	E_{ads} (kJ/mol)	$E_{surf/dep}$ (kJ/mol)	E_{surf} (kJ/mol)	E_{dep} (kJ/mol)
The mixed collector	−52.16	−7093.30	−6787.48	−253.66
DTAC	−35.10	−7110.72	−6787.66	−287.96

4. Conclusions

In this study, the monomineral flotation tests of hematite, quartz, and their artificially mixed minerals proved that the mixed DTAC/tall oil collector could improve the flotation separation of quartz from hematite. This application not only increased the efficiency of hematite reverse flotation but also strengthened the temperature adaptability of the collecting agents. The underlying mechanism was analyzed using surface tension tests, FT-IR, XPS, and MD simulations. The main conclusions are as follows:

1. The results of the monomineral flotation test revealed that, under the same conditions, the mixed collector surpassed the single DTAC. Quartz achieved a recovery rate of 94.67%, while hematite reached 8.69%, at a natural pH of around 7.0 and 298 K, indicating potential for the successful reverse flotation of hematite. In the comparative test, with artificially mixed minerals, the mixed collector exhibited strong selectivity compared to the single DTAC and displayed good temperature adaptability.
2. The surface tension tests confirmed the mixed collector's superior harvesting performance and low-temperature resistance, as well as the physical adsorption and hydrogen bonding of both single DTAC and mixed collector on quartz. It was observed that the adsorption of the mixed collector on the mineral surface was stronger when compared to that of the single DTAC.
3. MD simulations demonstrated that the mixed collector demonstrated more negative adsorption energy on the quartz surface than the single DTAC, leading to a firmer adsorption configuration conducive to flotation.

Author Contributions: Conceptualization, D.L. and J.W.; methodology, J.W.; formal analysis, X.H.; investigation, J.W. and X.H.; resources, D.L.; data curation, J.W.; writing—original draft preparation, J.W.; writing—review and editing, D.L. and J.W.; supervision, D.L. All authors have read and agreed to the published version of the manuscript.

Funding: This project was supported by the National Key Research and Development Program of China (2021YFC2903202) and the National Natural Science Foundation of China (52174270; 52174267; 51974366; 51674290).

Data Availability Statement: All data are contained within this article.

Acknowledgments: This work was supported, in part, by the High-performance Computing Center of Central South University.

Conflicts of Interest: The authors declare no conflicts of interest.

References

1. Iron Ore Statistics and Information | U.S. Geological Survey. Available online: https://www.usgs.gov/centers/national-minerals-information-center/iron-ore-statistics-and-information (accessed on 7 May 2024).
2. Liu, W.; Liu, W.; Dai, S.; Wang, B. Adsorption of bis(2-hydroxy-3-chloropropyl) dodecylamine on quartz surface and its implication on flotation. *Results Phys.* **2018**, *9*, 1096–1101. [CrossRef]
3. Yuan, S.; Xiao, H.; Wang, R.; Li, Y.; Gao, P. Improved iron recovery from low-grade iron ore by efficient suspension magnetization roasting and magnetic separation. *Miner. Eng.* **2022**, *186*, 107761. [CrossRef]
4. Li, W.; Liu, W.; Tong, K.; Liu, W.; Zhao, P.; Ding, R. Synthesis and flotation performance of a novel low-viscous cationic collector based on hematite reverse flotation desilication system. *Miner. Eng.* **2023**, *201*, 108190. [CrossRef]
5. Pattanaik, A.; Venugopal, R. Investigation of Adsorption Mechanism of Reagents (Surfactants) System and its Applicability in Iron Ore Flotation—An Overview. *Colloid Interface Sci. Commun.* **2018**, *25*, 41–65. [CrossRef]
6. Zhang, S.; Huang, Z.; Wang, H.; Liu, R.; Cheng, C.; Shuai, S.; Hu, Y.; Guo, Z.; Yu, X.; He, G.; et al. Flotation performance of a novel Gemini collector for kaolinite at low temperature. *Int. J. Min. Sci. Technol.* **2021**, *31*, 1145–1152. [CrossRef]
7. Zhang, X.; Gu, X.; Han, Y.; Parra-Álvarez, N.; Claremboux, V.; Kawatra, S.K. Flotation of Iron Ores: A Review. *Miner. Process. Extr. Metall. Rev.* **2021**, *42*, 184–212. [CrossRef]
8. Araujo, A.; Viana, P.; Peres, A. Reagents in iron ores flotation. *Miner. Eng.* **2005**, *18*, 219–224. [CrossRef]
9. Ding, Z.; Li, J.; Yuan, J.; Yu, A.; Wen, S.; Bai, S. Insights into the influence of calcium ions on the adsorption behavior of sodium oleate and its response to flotation of quartz: FT-IR, XPS and AMF studies. *Miner. Eng.* **2023**, *204*, 108437. [CrossRef]
10. Nakhaei, F.; Irannajad, M. Reagents types in flotation of iron oxide minerals: A review. *Miner. Process. Extr. Metall. Rev.* **2018**, *39*, 89–124. [CrossRef]

11. Filippov, L.; Severov, V.; Filippova, I. An overview of the beneficiation of iron ores via reverse cationic flotation. *Miner. Process.* **2014**, *127*, 62–69. [CrossRef]
12. Zhang, C.; Dai, H. Current Situation and Future Development of the Collector for Reverse Flotation of Iron Ore. *Multipurp. Util. Miner. Resour.* **2012**, *2*, 3–6.
13. Liu, W.; Tong, K.; Ding, R.; Liu, W.; Zhao, P.; Sun, W.; Zhao, Q.; Zhao, S. Synthesis of a novel hydroxyl quaternary ammonium collector and its selective flotation separation of quartz from hematite. *Miner. Eng.* **2023**, *200*, 108109. [CrossRef]
14. A, Vidyadhar; N, Kumari; R, P, Bhagat; Adsorption mechanism of mixed collector systems on hematite flotation. *Miner. Eng.* **2012**, *26*, 102–104. [CrossRef]
15. El Kadi, N.; Martins, F.; Clausse, D.; Schulz, P. Critical micelle concentrations of aqueous hexadecytrimethylammonium bromide-sodium oleate mixtures. *Colloid Polym.* **2003**, *281*, 353–362. [CrossRef]
16. Özün, S.; Ulutaş, Ş. Interfacial Behavior of Anionic/Cationic Flotation Collectors in Mixed Aqueous Solutions and Their Effect on Flotation Recovery of Quartz. *J. Surfactants Deterg.* **2019**, *22*, 61–71. [CrossRef]
17. Yan, W.; Liu, C.; Feng, Q.; Zhang, W. Flotation separation of scheelite from calcite using mixed collectors. *Int. J. Miner. Process.* **2017**, *169*, 106–110. [CrossRef]
18. Tian, J.; Xu, L.; Deng, W.; Jiang, H.; Gao, Z.; Hu, Y. Adsorption mechanism of new mixed anionic/cationic collectors in a spodumene-feldspar flotation system. *Chem. Eng. Sci.* **2017**, *164*, 99–107. [CrossRef]
19. Bai, S.; Li, J.; Bi, Y.; Yuan, J.; Wen, S.; Zhan, D. Adsorption of sodium oleate at the microfine hematite/aqueous solution interface and its consequences for flotation. *Int. J. Min. Sci. Technol.* **2023**, *33*, 105–113. [CrossRef]
20. Ozun, S.; Atalay, M.U. A comparative study on interactions of ionic collectors with orthoclase. *Physicochem. Probl. Miner. Process.* **2016**, *52*, 955–972.
21. Cheng, Z.; Zhu, Y.; Li, Y.; Butt, S. Experimental and MD simulation of 3-dodecyloxypropanamine and 3-tetradecyloxypropylamine adsorbed onto quartz (1 0 1) surface. *Int. J. Min. Sci. Technol.* **2021**, *31*, 1033–1042. [CrossRef]
22. Liu, A.; Fan, J.; Fan, M. Quantum chemical calculations and molecular dynamics simulations of amine collector adsorption on quartz (0 0 1) surface in the aqueous solution. *Int. J. Miner. Process.* **2015**, *134*, 1–10. [CrossRef]
23. Li, L.; Hao, H.; Yuan, Z.; Liu, J. Molecular dynamics simulation of siderite-hematite-quartz flotation with sodium oleate. *Appl. Surf. Sci.* **2017**, *419*, 557–563. [CrossRef]
24. Wang, X.; Liu, W.; Duan, H.; Wang, B.; Han, C.; Wei, D. The adsorption mechanism of calcium ion on quartz (1 0 1) surface: A DFT study. *Powder Technol.* **2018**, *329*, 158–166. [CrossRef]
25. Zhu, Y.; Luo, B.; Sun, C.; Lun, J.; Sun, H.; Li, Y.; Han, Y. Density functional theory study of α-Bromolauric acid adsorption on the α-quartz (1 0 1) surface. *Miner. Eng.* **2016**, *92*, 72–77. [CrossRef]
26. Quezada, G.; Piceros, E.; Saavedra, J.; Pedro, R.; Ricardo, I. Polymer affinity with quartz (1 0 1) surface in saline solutions: A molecular dynamics study. *Miner. Eng.* **2022**, *186*, 107750. [CrossRef]
27. Zhang, S.; Ren, J.; Wen, Y.; Shao, X. Flotation separation of diaspore and kaolinite by using dodecanamide and cetyltrimethylammonium bromide as an effective mixed collector. *Miner. Eng.* **2022**, *179*, 107434. [CrossRef]
28. Zhou, Y.; Zhang, Y.; Li, P.; Li, G.; Jiang, T. Comparative study on the adsorption interactions of humic acid onto natural magnetite, hematite and quartz: Effect of initial HA concentration. *Powder Technol.* **2014**, *251*, 1–8. [CrossRef]

minerals

MDPI

Article

Inhibiting Mechanism of High pH on Molybdenite Flotation. An Experimental and DFT Study

Enxiang Wang [1,2], He Wan [1,3,*], Juanping Qu [4,*], Peng Yi [1] and Xianzhong Bu [1]

1 School of Resources Engineering, Xi'an University of Architecture and Technology, Xi'an 710055, China
2 JinDuiCheng Molybdenum Group Co., Ltd., Weinan 714000, China
3 Lithium Resources and Lithium Materials Key Laboratory of Sichuan Province, Tianqi Lithium Corporation, Chengdu 610000, China
4 Oulu Mining School, University of Oulu, FI-90014 Oulu, Finland
* Correspondence: wanhe@xauat.edu.cn (H.W.); juanping.qu@student.oulu.fi (J.Q.); Tel.: +86-02982203408 (H.W.)

Abstract: The inhibiting mechanism of high pH on the molybdenite flotation was studied using an experimental and DFT method. The experimental results found that adverse effects of pH on molybdenite flotation should be attributed to the adsorption of OH^- on molybdenite [100] surface (MS100). The DFT results show the adsorption energy of H_2O/OH^- to molybdenite [001] surface (MS001) and MS100 is $-8.61/288.30$ kJ·mol^{-1} and $-226.81/-302.44$ kJ·mol^{-1}. These indicate that H_2O is weakly adsorbed on MS001, while OH^- is not. Both H_2O and OH^- can be adsorbed onto MS100. The adsorption energy of OH^- to MS100 is much stronger than that of H_2O. The results of state density and charge transfer of the adsorption of OH^- on MS100 further show that OH^- can be chemically adsorbed on MS100 through the bonding of the O atom of OH^- and the Mo atom of MS100. This causes a significant reduction in the MS100 hydrophobicity and deteriorates the fine molybdenite flotation.

Keywords: high pH; molybdenite flotation; inhibiting mechanism; density functional theory

1. Introduction

pH is one of the essential affecting factors in the flotation process [1]. The flotation process can be carried out more efficiently with the appropriate pH. Park et al. showed that with increasing pH of the pulp, the flotation recovery of molybdenite continues to decrease [2]. Qiu and You also suggested that the pulp's pH will influence the collector's adsorption behavior on the molybdenite surface, thus affecting its flotation performance [3,4]. However, in the absence of a collector, the flotation recovery of molybdenite still continuously decreases as the pulp's pH increases [5]. This indicates that the inhibitory role of high pH on molybdenite flotation does not only reduce the adsorption behavior of the collectors on the molybdenite surface but also changes the surface properties of the molybdenite.

Lin and Hao found that these changes in the molybdenite surface properties are mainly reflected in the increasing electronegativity and deteriorating hydrophobicity of the molybdenite particles as the pulp pH increases [6,7]. They speculated that alkaline conditions may improve the oxidation performance of molybdenite surfaces and cause molybdenite particles to be more hydrophilic and have lower floatability. However, the mechanism of the effect of pH on the molybdenite flotation under collectorless conditions has not been reported. From another point of view, pH also represents the hydroxide ion (OH^-) concentration, indicating that OH^- concentration in pulp may also affect molybdenite flotation.

Molybdenite belongs to a hexagonal crystal system and has a typical layered structure [8,9]. The surface anisotropy of molybdenite on a macroscopic scale has been known for a long time [10,11]. This means that the molybdenite [001] surface (MS001) is a nonpolar surface with strong hydrophobicity, while MS100 is a polar surface with strong hydrophilicity [12]. The difference between MS001 and MS100 is caused by the fact that MS001 breaks molecular bonds while MS100 breaks covalent bonds during the grinding process [13]. Therefore, the adsorption behavior of OH^- on the MS001 and MS100 is bound to differ. Moreover, the adsorption mechanism of OH^- on the MS001/MS100 has not been reported.

In this paper, the effect of pulp pH on molybdenite flotation is first investigated by zeta potential and contact angle, after which the inhibiting mechanism of high pH on MS001/MS100 is analyzed by the density functional theory (DFT). The adsorption energies, bond lengths, Mulliken bond population, and electronic properties (including difference densities, the density of states, and Mulliken charge population) of H_2O and OH^- on MS001/MS100 are calculated. The competitive adsorption behaviors between H_2O and OH^- on MS001/MS100 are discussed. Finally, the inhibiting mechanism of OH^- on the MS001/MS100 is analyzed at the molecular and atomic levels.

2. Materials and Methods

2.1. Materials and Reagents

The pure molybdenite sample (PMS) used in this study was purchased from Huadong Ye's Stone Specimen Company in Guangdong, China. A portion of bulk molybdenite crystals with a complete crystal structure was used to prepare contact angle test samples. They were flaked and polished to obtain the sample of MS001 and MS100, respectively (the MS001 and MS100 are shown in Figure 1).

Figure 1. The pure molybdenite sample. (**a**) MS001; (**b**) MS100.

The remaining molybdenite crystals were used to prepare samples for flotation and zeta potential tests. They were first crushed and ground in a porcelain ball mill. Then, the PMS was screened to obtain the −0.038 mm fraction for tests and analyses. This material assayed 59 wt% Mo. The XRD result of the purified molybdenite sample shows 99 wt% molybdenite, as shown in Figure 2.

Pine oil was used as frother in molybdenite flotation. Sodium hydroxide (NaOH) and hydrochloric acid (HCl) were used to adjust the pH of the water. Synthetic water was used in the experiment. It was prepared by dissolving NaOH and HCl in distilled water.

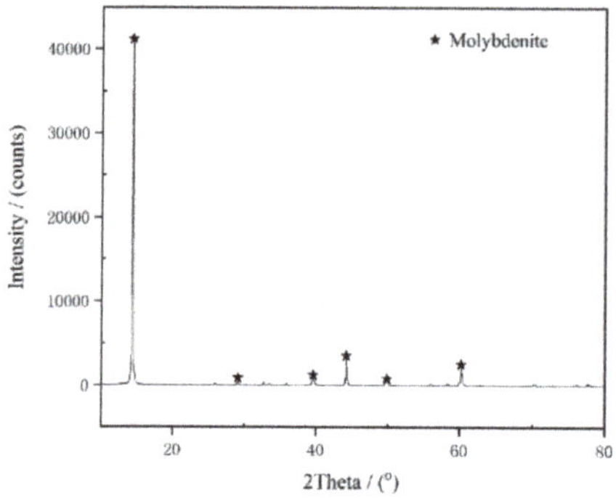

Figure 2. The XRD result of the PMS.

2.2. Flotation Tests

PMS flotation was conducted in an RK/FGC 40 mL hitch groove flotation cell at a rotating speed of 1750 r/min. In each set of experiments, 2.0 g of −0.038 mm PMS was dispersed in 40 mL synthetic water with different pH values and reacted for 2 min by magnetic stirring. Then, it was poured into a flotation machine, and a certain amount of frother (25 mg/L for all flotation tests) was added, and the flotation time was 3 min. The product was collected, filtered, dried, and weighed. Each test was repeated five times, and the average flotation recovery was reported.

2.3. Zeta Potential Measurements

Zeta potential was measured with a zeta potential analyzer (Nano plus, Micromeritics Instrument Corp., USA). The suspension containing less than 5% PMS was conditioned for 5 min in a solution of pH = 4~12. The suspension was then transferred to a sample vessel for the zeta potential measurement at room temperature (20 °C). The measurement was repeated five times, and the average zeta potential was reported.

2.4. Contact Angle Measurements

The contact angle was measured using a contact angle tester (JY-82C, Chengde Dingsheng Testing Equipment Co Ltd., Chengde, China). The sample of MS001 and MS100 was soaked for 10 min in a solution of pH = 4~12. The sample was then transferred to the viewing platform. Finally, one drop of water was dropped onto the sample's surface for the contact angle measurement at room temperature (20 °C). The measurement was repeated five times, and the average contact angle was reported.

2.5. Simulation Details

The adsorption model of water molecules and OH^-/H^+ on the molybdenite surface (white is the H atom, red is the O atom) is shown in Figure 3. All density functional theory (DFT) calculations were regulated using the Cambridge Serial Total Energy Package (CASTEP) Module in Materials Studio 6.0. The simulation details are consistent with the simulation process mentioned in our published article on minerals engineering [14]; moreover, it can also be found in the Supplementary Materials.

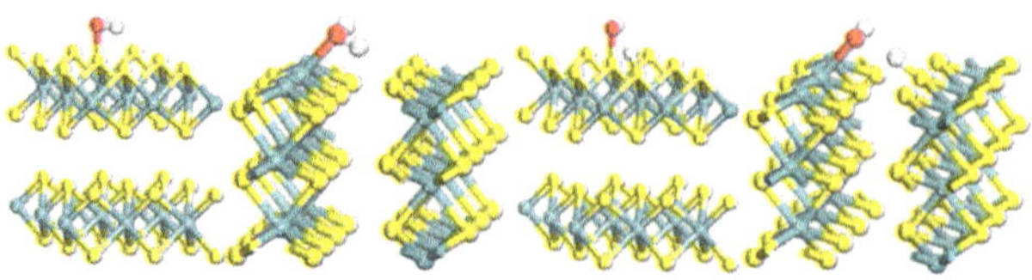

Figure 3. Adsorption model of water molecules and OH^-/H^+ on molybdenite surface (white is H atom, red is O atom).

3. Results and Discussion

3.1. Flotation Test

The effect of pH on the flotation of molybdenite was studied, and the results are shown in Figure 4. As can be seen in Figure 4, with the increase in the pulp pH, the recovery curve of molybdenite flotation in the absence of collector tends to be flat and then decreasing. The result is consistent with the results of Chander et al. in the presence of a collector [15]. These indicate that high pH has an inhibiting effect on molybdenite flotation regardless of the presence or absence of a collector. It is well known that high pH equates to high OH^- concentrations. Therefore, there may be significant OH^- adsorbed on the molybdenite surface at high pH pulp. It can cause a reduction in the floatability of the molybdenite and thus deteriorate the molybdenite flotation.

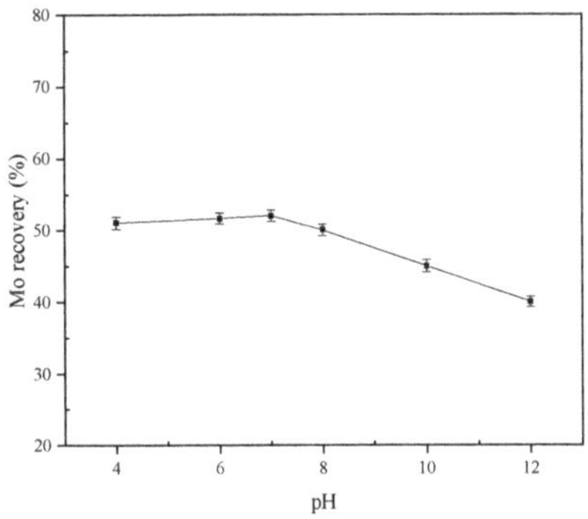

Figure 4. The effect of pH on the flotation of molybdenite.

3.2. Zeta Potential Measurements

The effect of pH on the zeta potential of molybdenite is shown in Figure 5. As the pH of the pulp increases, the zeta potential of the molybdenite surface tends to decrease continuously. This result supports the view that the adsorption of OH^- on the molybdenite surface causes a deterioration in the flotation of molybdenite. Moreover, this is also consistent with the Eh–pH diagram for molybdenum in a Mo–H_2O system [16]. However, the decrease in the zeta potential of the molybdenite caused by an increase in pH does not indicate that it caused a decrease in the molybdenite floatability.

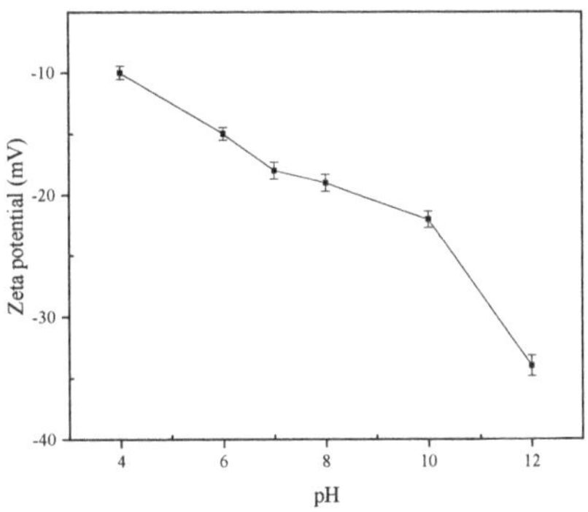

Figure 5. The effect of pH on the zeta potential of molybdenite.

3.3. Contact Angle Measurements

The contact angle of MS001/MS100 with pH ≥ 7 is shown in Figure 6.

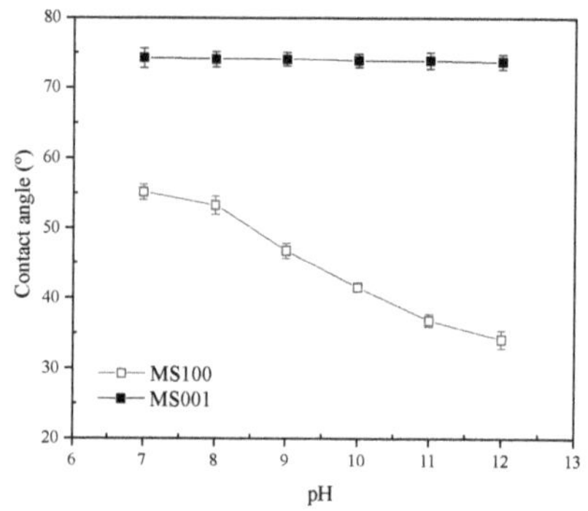

Figure 6. Contact angle of MS001/MS100 with pH $\geq$ 7.

The results of Figure 6 found that as the pH of the pulp increases, the contact angle of MS001 tends to remain constant, whereas the contact angle of MS100 tends to decrease. This indicates that alkaline conditions have no significant effect on the hydrophobicity of MS001 while causing a decrease in the hydrophobicity of MS100. Therefore, the inhibitory role of high pH on molybdenite flotation may be due to the changing of the MS100 properties. This result is in agreement with the anisotropies of the molybdenite surface [17]. In other words, MS001 is a nonpolar surface with stable surface properties; MS100 is a polar surface with unstable surface properties. The alkaline conditions are favorable to the changing of the MS100 properties. This is the reason for the more hydrophilic and lower floatability of MS100 in alkaline conditions.

3.4. *Inhibiting Mechanism of High pH on Molybdenite Flotation*

To further study the inhibiting mechanism of high pH on molybdenite flotation, DFT was used to analyze the interaction of water (H_2O) and hydroxide/hydrogen ions (OH^-/H^+) with the molybdenite surface at the atomic and electronic levels.

3.4.1. Adsorption Energy of H_2O and OH^-/H^+ on MS001/ MS100

The adsorption energies of H_2O and OH^-/H^+ on MS001/MS100 are shown in Table 1.

Table 1. The adsorption energy of H_2O and OH^-/H^+ with MS001/MS100.

Adsorbate	Surface	$E_{surface+ion}$/kJ·mol^{-1}	E_{ion}/kJ·mol^{-1}	$E_{surface}$/kJ·mol^{-1}	ΔE/kJ·mol^{-1}
H_2O	001	46,312.59	471.93	5093.40	−8.61
	100	46,286.81	471.93	15,270.84	−226.81
OH^-/H^+	001	46,309.51	471.93	5093.40	288.30
	100	46,287.59	471.93	15,270.84	−302.44

The results in Table 1 show that the interaction energies of H_2O with MS001/MS100 are −8.61/−226.81kJ·mol^{-1}, respectively. The interaction energies of OH^-/H^+ with MS001/MS100 are 288.30/−302.44kJ·mol^{-1}, respectively. The adsorption energy shows that MS001 is strongly hydrophobic and MS100 is strongly hydrophilic. According to the Mulliken charge population of the atoms on the MS001 and MS100 (0.2 negative charges on the S-plane and 0.12 positive charge on the Mo-plane), OH^- is easily adsorbed on the Mo-plane, and H^+ is easily adsorbed on the S-plane. Therefore, OH^-/H^+ is more easily adsorbed on MS100 than water molecules, enhancing its hydrophilicity.

3.4.2. Electronic Properties of H_2O and OH^-/H^+ at MS001/MS100

The electronic properties of H_2O and OH^-/H^+ at MS001/MS100 can reveal their adsorption mechanism on molybdenite surfaces at the molecular and electronic levels. It mainly includes the bond lengths, Mulliken bond population, differential density, density of states, and Mulliken charge population.

(1) Analysis of Water Bonding on the Surface of Molybdenite

The stable adsorption configurations of H_2O in MS001/MS100 are shown in Figures 7 and 8, respectively. To quantitatively characterize the adsorption configuration of H_2O on the MS001/MS100, the Mulliken bond populations and their corresponding bond lengths after stable adsorption of H_2O on the MS001/MS100 are listed in Table 2.

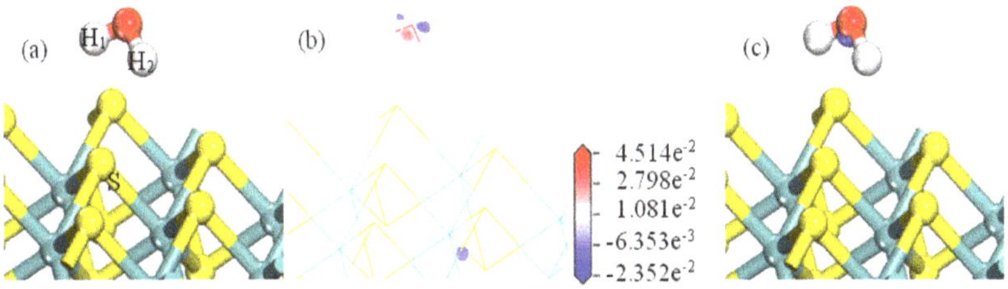

Figure 7. Differential charge density diagram of H_2O after adsorption in MS001 ((**a**): adsorption configuration, (**b**): differential charge density section, (**c**): differential charge density diagram).

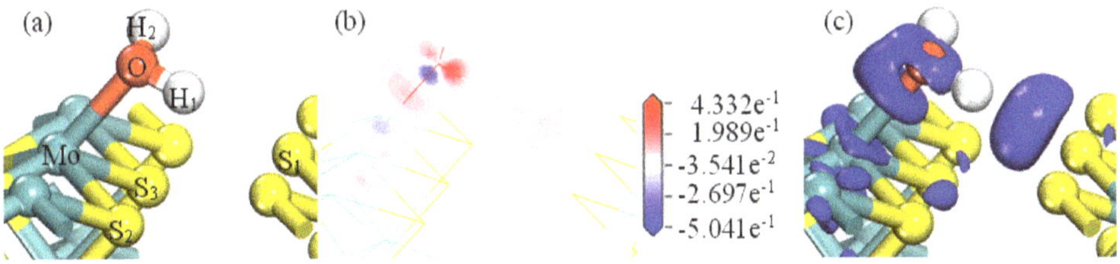

Figure 8. Differential charge density diagram of H_2O after adsorption in MS100 ((**a**): adsorption configuration, (**b**): differential charge density section, (**c**): differential charge density diagram).

Table 2. The adsorption energy of H_2O and OH^-/H^+ with MS001/MS100.

The Adsorption Configuration	Bond	Population	Length/(nm)
H_2O–MS001	H_1-S	0.00	2.94672
	H_2-S	0.00	2.93816
H_2O–MS100	O-Mo	0.10	2.25993
	H_1-Mo	0.16	2.70418
	H_2-Mo	0.17	2.78119
	H_1-S_1	0.09	2.34405
	H_1-S_2	0.02	2.76602
	H_1-S_3	0.01	2.98579

As shown in Figure 7a, when the H_2O is adsorbed on the MS001, the water molecules are far away from the MS001. As shown in Figure 7b, there is essentially no electron cloud density between the hydrogen atom and the S atom of MS001. Figure 7c shows that there are essentially no electrons around the S atom of MS001. There is only weak electron aggregation between the O and H atoms in the water molecule. Combining the results in Table 2 shows that the H atoms of the water molecules are weakly adsorbed to the S atoms on MS001. The corresponding bond lengths are 2.94672 nm and 2.93816 nm, respectively. The corresponding Mulliken bond population is zero. This is consistent with the results for the adsorption energy of water with the MS001 in Table 1 ($\Delta E = -8.61$ kJ.mol^{-1}).

Figure 8a shows that when water molecules are adsorbed on the MS100, the water molecules are essentially located between the Mo-plane and the S-plane of MS100. Combining the results in Table 2 shows that the H atoms of H_2O are located almost between the Mo and S surfaces of the MS100. The O atom of H_2O may be bonded to the Mo atom on MS100. H_1 forms a strong hydrogen bonding interaction with S_1 (with a sizeable covalent population). The corresponding bond lengths are 2.25993 nm and 2.34405 nm. The corresponding Mulliken bond population value is 0.10 and 0.09, respectively.

As shown in Figure 8b, there is a high electron cloud density between the O and H atoms of the water molecule and the Mo atoms of the MS100. The corresponding population values are −0.16 and −0.17. A high charge density also exists between the H_1 atom and the S_1 atom.

Figure 8c shows that the electron cloud is located between the water molecules and the MS100. The electron aggregation occurs between the Mo atoms of MS100 and the O atoms in the water and between the hydrogen bonds formed by the H_1 and S_1 atoms. These indicate that the O/H atoms of the water molecule interact with the Mo/S atoms of MS100, respectively.

(2) Analysis of OH^-/H^+ Bonding on the MS100

The results of Table 1 show that the interaction energies of OH^-/H^+ with MS001 are 288.30kJ·mol^{-1}. This indicates that OH^-/H^+ exhibits no adsorption on MS001. Therefore, only OH^-/H^+ bonding on the MS100 needs to be analyzed.

The Mulliken bond populations and their corresponding bond lengths after stable adsorption of OH^-/H^+ on the MS100 are listed in Table 3 and Figure 9.

Table 3. Mulliken population and bond length of OH^-/H^+ adsorbed on molybdenite surface.

The Adsorption Configuration	Bond	Population	Length/(nm)
	O-Mo	0.44	1.90191
	H_1-Mo	0.27	2.62442
OH^-/H^+–MS100	H_2-S_1	0.72	1.36465
	H_2-S_2	0.01	2.60470
	H_2-S_3	0.00	2.59003

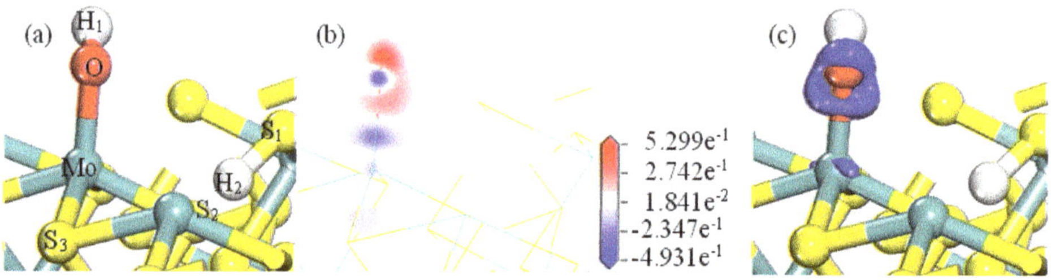

Figure 9. Differential charge density diagram of OH^-/H^+ after adsorption in MS100 ((**a**): adsorption configuration, (**b**): differential charge density section, (**c**): differential charge density diagram).

Figure 9a shows that OH^- and H^+ can adsorb to the MS100 by bonding to the Mo and S atoms on MS100, respectively. Figure 9b shows a clear charge transfer when the O atom of OH^- is bonded to the Mo atom on the MS100. The Mo atom loses its charge point, and the O atom in OH^- gains electrons. It is evident from Figure 9c that the O atom in OH^- is surrounded by a cloud of electrons. The electrostatic adsorption between the OH^- and Mo atoms of MS100 enhances the adsorption of OH^- to the Mo-plane of MS100. The S atoms of MS100 exhibit a negative charge, which favors the adsorption of H^+ or cations on the S-plane of MS100. The results in Table 3 further show that the O atom in OH^- can strongly bond with the Mo atom of the Mo-plane, with a population value of 0.44 and bond lengths of 1.90191 nm. The H_2 atom can also strongly bond with the S_1 atom of the S-plane, with a population value of 0.72 and bond lengths of 1.36465 nm. These indicate that strong adsorption of OH^-/H^+ occurs with MS100. This is consistent with the calculation results of adsorption energy ($\Delta E(100)\text{-}OH^-/H^+$) = −302.44 kJ·mol^{-1}, showing the strong hydrophilicity of MS100.

(3) State Density and Charge Transfer of H_2O Adsorption Process

To study the variation of density of states and charge transfer during the adsorption of H_2O on the MS001/MS100, the S atom on MS001 and the H atom in H_2O molecules were selected as the research objects. The S atom on the S-plane of MS100 (MS100-S_1 in the differential density map) and the H atom in the water molecule (H_1) were selected as the research objects (bond population 0.09, interatom distance 2.34405 nm) to analyze the changes in state density and bonding before and after adsorption. Figure 10 shows the density of states before and after the interaction of the H atoms in water with the S atoms on the molybdenite surface.

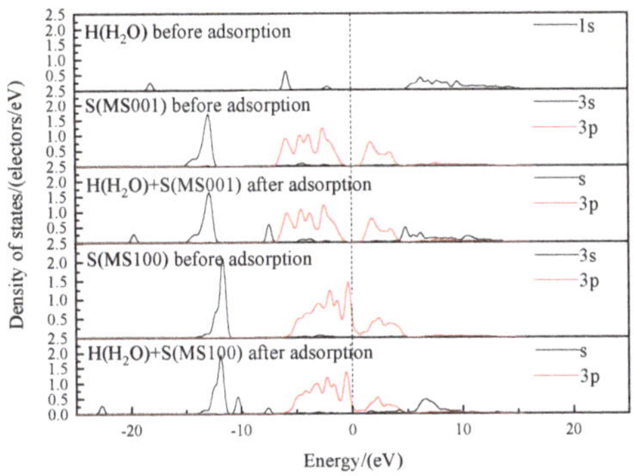

Figure 10. The variation of density of states during the adsorption of H_2O on the MS001/MS100.

As shown in Figure 10, after the adsorption of water molecules on MS100, the H atoms interact with the S atoms, causing a decrease in the density of electronic states near their Fermi energy levels. The overall density of states moves towards lower energy levels. This indicates that the energy of the electron orbitals decreases after the adsorption of water molecules on the MS100. Then, the system becomes stable. Meanwhile, between −7.5 and −7.1 eV are the bonding interactions between the H and S atoms, and between 5 and 8 are their antibonding interactions. Both of these interactions are weak, but the bonding interaction is stronger than the antibonding interaction. Therefore, the interaction between water molecules and molybdenite is in the form of a bonding interaction, with a bonding population of 0.09.

Tables 4 and 5 show the Mulliken charges of the bonding atoms involved before and after water adsorption on the MS001/MS100, respectively.

Table 4. The charge transfer during the adsorption of H_2O on the MS001.

Atomic	State	s	p	d	Population	Charge/(e)
H_1	Before	0.47	0.00	0.00	0.47	0.53
	After	0.51	0.00	0.00	0.51	0.49
H_2	Before	0.47	0.00	0.00	0.47	0.53
	After	0.51	0.00	0.00	0.51	0.49
S	Before	1.86	4.16	0.00	6.02	0.02
	After	1.85	4.18	0.00	6.03	0.03

Table 5. The charge transfer during the adsorption of H_2O on the MS100.

Atomic	State	s	p	d	Population	Charge/(e)
O	Before	1.89	5.18	0.00	7.07	1.07
	After	1.85	4.99	0.00	6.83	0.83
H_1	Before	0.47	0.00	0.00	0.47	0.53
	After	0.57	0.00	0.00	0.57	0.43
H_2	Before	0.47	0.00	0.00	0.47	0.53
	After	0.54	0.00	0.00	0.54	0.46
S_1	Before	1.88	4.32	0.00	6.20	0.20
	After	1.86	4.42	0.00	6.28	0.28
S_2	Before	1.86	4.17	0.00	6.03	0.03
	After	1.85	4.17	0.00	6.03	0.03
S_3	Before	1.86	4.17	0.00	6.03	0.03
	After	1.85	4.15	0.00	6.00	0.00
Mo	Before	2.44	6.26	5.12	13.82	0.18
	After	2.43	6.29	5.04	13.76	0.24

As shown in Table 4, when the water molecules are adsorbed on the MS001, the density of states of the S atoms on the MS001 remains essentially unchanged. There is no significant charge transfer from the MS001. Combined with the results in Figure 10, there is no charge transfer occurring in the vicinity of the O atom in the water molecule by aggregating only a small number of electrons. This suggests that water molecules do not adsorb stably or weakly on the MS001.

Table 5 shows that the 1s orbital population of the H_1 atom in the water molecule increases by 0.1, and the positive charge decreases by 0.1 e. The 3s orbital population of the S_1 atom on the S-plane of MS100 decreases by 0.02, and the 3d orbital population increases by 0.1, increasing the total charge by 0.08 e. This indicates that the H_1 atom loses electrons, and the S_1 atom gains electrons. The result is consistent with the results of Figure 10. These indicate that water molecules can undergo stable adsorption on the MS100.

(4) State Density and Charge Transfer of OH^- Adsorption Process

The results of Table 1, Figure 8, and Figure 9 show that OH^-/H^+ exhibits no adsorption on MS001. Therefore, only the adsorption of OH^- on MS100 needs to be investigated. The O atom in OH^- and the Mo atom on the Mo-plane of MS100 were selected to analyze the density of states of the atoms before and after adsorption. The results are shown in Figure 11.

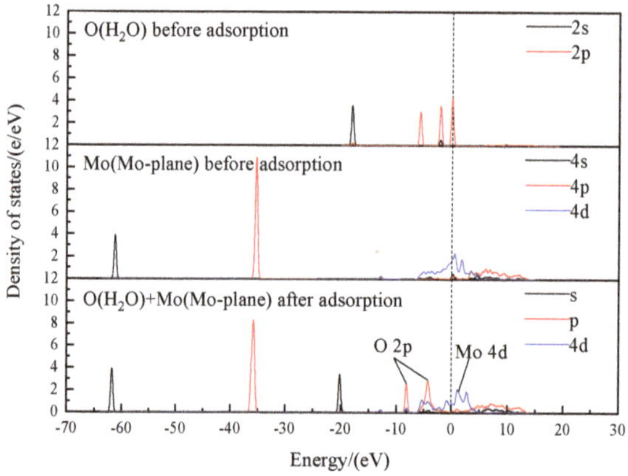

Figure 11. The variation of density of states during the adsorption of OH^- on the MS001/MS100.

As shown in Figure 11, both the 4d orbital of the Mo atom and the 2p orbital of the O atom have a high density of states near the Fermi energy level. This indicates that both are highly active. After the interaction between the Mo atom and the O atom, the density of states of the 2p orbital of the O atom decreases and moves to the lower energy level, while the density of states of the 4p and 4d orbitals of the Mo atom decreases and the 4d orbital moves to the higher energy level, resulting in increased antibonding. The intervals of −8.6 to −7.5 eV/−6.1 to −2.5 eV are the bonding interactions between the 2p orbital of the O atom and the 4s/4d orbitals of the Mo atom, respectively.

The OH^- ions can adsorb onto the Mo-plane of MS100 by chemisorption and form a hydroxy-molybdenum compound with the Mo atoms. It will cause a drop in the potential of the MS100. This result is consistent with the effect of OH^- ion concentration on the zeta potential of molybdenite particles.

To obtain the charge transfer of atoms during the adsorption of OH^- on MS100, the Mulliken charge population of the directly bonded atoms was analyzed. The results of the analytical calculations are shown in Table 6.

Table 6. The charge transfer during the adsorption of OH^- on the MS100.

Atomic	State	s	p	d	Population	Charge/(e)
O	Before	1.89	5.18	0.00	7.07	1.07
	After	1.86	4.93	0.00	6.78	0.78
H_1	Before	0.47	0.00	0.00	0.47	0.53
	After	0.57	0.00	0.00	0.57	0.43
H_2	Before	0.47	0.00	0.00	0.47	0.53
	After	0.95	0.00	0.00	0.95	0.05
S_1	Before	1.89	4.32	0.00	6.20	0.20
	After	1.85	4.22	0.00	6.08	0.08
S_2	Before	1.86	4.17	0.00	6.03	0.03
	After	1.86	4.20	0.00	6.05	0.05
S_3	Before	1.86	4.17	0.00	6.03	0.03
	After	1.85	4.23	0.00	6.09	0.09
Mo	Before	2.44	6.26	5.12	13.82	0.18
	After	2.43	6.32	4.82	13.57	0.43

As shown in Table 6, the 1s state of the H atom in OH^- gains 0.1e, and the 2p state of the O atom loses 0.25e. The 4s state of the Mo atom loses 0.01e; the 4d state loses 0.3e; the 4p state gains 0.06e; the Mo atom loses a total of 0.25e. This indicates that in the process of OH^- adsorption on MS100, in addition to the formation of hydrogen bonds between the H atoms and the S atoms of MS100, there may also be electrostatic adsorption when the OH^- is near the Mo-plane of MS100.

4. Conclusions

This paper studied the effect of pH on the flotation of molybdenite by an experimental and DFT method at the molecular and atomic levels and discussed the effect mechanism. The main conclusions are summarized as follows:

(1) With increasing pH of the pulp, the curve of molybdenite flotation recovery first tends to flatten out and then gradually decreases after pH above 7, the curve of the zeta potential of the molybdenite surface tends to decrease continuously, and the contact angle of MS100 tends to increase and then decrease. These indicate that the effect of pH on molybdenite flotation should be attributed to the adsorption behavior of the OH^- on MS100. Both H_2O and OH^- can be adsorbed onto the MS100.

(2) The adsorption energy of the H_2O/OH^- to the MS001 and MS100 is −8.61/288.30 kJ·mol−1 and −226.81/−302.44 kJ·mol^{-1}. This indicates that H_2O is weakly adsorbed on the MS001, while OH^- is not. The adsorption energy of the OH^- to the MS100 is much stronger than that of H_2O.

(3) The adsorption of H_2O on MS100 is through electrostatic interaction/hydrogen bond between the O/H atom of H_2O and the Mo/S atom of MS100. However, the adsorption of OH^- on MS100 is achieved in two steps: the OH^- ion is physically adsorbed on the MS100 by electrostatic interaction/hydrogen bonding between the O/H atom of OH^- and the Mo/S atom of MS100; then, the OH^- ion is further chemically adsorbed on the MS100 through the bonding of the O atom of OH^- and the Mo atom of MS100, and reacts to form a hydroxy-molybdenum compound. This causes a significant reduction in the MS100 hydrophobicity and deteriorates the fine molybdenite flotation.

Supplementary Materials: The following supporting information can be downloaded at: https://www.mdpi.com/article/10.3390/min14070663/s1, Figure S1: Single crystal cell structure of molybdenite; Figure S2: The convergence test of cut-off energy; Figure S3: The convergence test of k-point density; Figure S4: The effect of molybdenite atomic layer number on the surface energy of the MS001/MS100; Figure S5: The structure of the extended MS001; Figure S6: Adsorption model of water molecules and OH^-/H^+ on molybdenite surface.

Author Contributions: Conceptualization, H.W. and E.W.; methodology, J.Q., E.W. and P.Y.; software, P.Y.; validation, P.Y. and J.Q.; formal analysis, H.W. and P.Y.; investigation, H.W., E.W. and P.Y.; resources, E.W.; data curation, J.Q. and P.Y.; writing—original draft preparation, H.W. and P.Y.; writing—review and editing, E.W. and J.Q.; visualization, P.Y. and X.B.; supervision, H.W. and X.B.; project administration, H.W.; funding acquisition, H.W. All authors have read and agreed to the published version of the manuscript.

Funding: This research was funded by the Natural Science Foundation of China (Grant No. 52274271), the Lithium Resources and Lithium Materials Key Laboratory of Sichuan Province (Grant No. LRMKF202310).

Data Availability Statement: The data presented are available in the article.

Acknowledgments: Northeastern University contributed Materials Studio Software; Saija Luukkanen of the University of Oulu contributed to the writing review.

Conflicts of Interest: Enxiang Wang is an employee of JinDuiCheng Molybdenum Group Co., Ltd. The paper reflects the views of the scientists and not the company.

References

1. Liu, W.; Moran, C.J.; Vink, S. A review of the effect of water quality on flotation. *Miner. Eng.* **2013**, *53*, 91–100. [CrossRef]
2. Park, C.H.; Jeon, H.S. The effect of sodium silicate as pH modifier and depressant in the froth flotation of molybdenite ores. *Mater. Trans.* **2010**, *51*, 1367–1369. [CrossRef]
3. Qin, W.; Wu, J.; Jiao, F.; Zeng, J. Mechanism study on flotation separation of molybdenite from chalcocite using thioglycollic acid as depressant. *Int. J. Min. Sci. Technol.* **2017**, *27*, 1043–1049. [CrossRef]
4. You, X.; Li, L.; Lyu, X. Flotation of molybdenite in the presence of microemulsified collector. *Physicochem. Probl. Miner. Process.* **2017**, *53*, 333–341.
5. Zhang, W.; Tao, L.; Xun, L.; Qi, Z.; Pooley, S.; Sun, W.; Gao, Z. Improved flotation of molybdenite from talc using a selective reagent scheme. *Miner. Eng.* **2022**, *176*, 107324. [CrossRef]
6. Lin, S.; Wang, C.; Liu, R.; Sun, W.; Jing, G. Surface characterization of molybdenite, bismuthinite, and pyrite to identify the influence of pH on the mineral floatability. *Appl. Surf. Sci.* **2022**, *577*, 151756. [CrossRef]
7. Hao, J.; Liu, J.; Yu, Y.; Gao, H.; Qin, X.; Bai, X. Depressants for separation of chalcopyrite and molybdenite: Review and prospects. *Miner. Eng.* **2023**, *201*, 108209. [CrossRef]
8. McClung, C.R. Molybdenite polytypism and its implications for processing and recovery: A geometallurgical-based case study from Bingham Canyon Mine, Utah. *Miner. Metall. Process.* **2016**, *33*, 149–154. [CrossRef]
9. Dickinson, R.G.; Pauling, L. The crystal structure of molybdenite. *J. Am. Chem. Soc.* **1923**, *45*, 1466–1471. [CrossRef]
10. Lu, Z.; Liu, Q.; Xu, Z.; Zeng, H. Probing Anisotropic Surface Properties of Molybdenite by Direct Force Measurements. *Langmuir* **2015**, *31*, 11409–11418. [CrossRef] [PubMed]
11. Bentley, C.L.; Kang, M.; Maddar, F.M.; Li, F.; Walker, M.; Zhang, J.; Unwin, P.R. Electrochemical maps and movies of the hydrogen evolution reaction on natural crystals of molybdenite (MoS2): Basal vs. edge plane activity. *Chem. Sci.* **2017**, *8*, 6583–6593. [CrossRef] [PubMed]
12. Castro, S.; Lopez-Valdivieso, A.; Laskowski, J.S. Review of the flotation of molybdenite. Part I: Surface properties and floatability. *Int. J. Miner. Process.* **2016**, *148*, 48–58. [CrossRef]

13. Yang, B.; Song, S.; Lopez-Valdivieso, A. Effect of particle size on the contact angle of molybdenite powders. *Min. Proc. Ext. Met. Rev.* **2014**, *35*, 208–215. [CrossRef]
14. Wan, H.; Yi, P.; Song, X.; Luukkanen, S.; Qu, J.; Yang, W.; Bu, X. Role of improving molybdenite flotation by using aromatic hydrocarbon collector in high-calcium water: A multiscale investigation. *Miner. Eng.* **2023**, *191*, 107984. [CrossRef]
15. Chander, S.; Hogg, R.; Fuerstenau, D.W. Characterization of the wetting and dewetting behavior of powders. *KONA Powder Part. J.* **2007**, *25*, 56–75. [CrossRef]
16. Ochs, M.; Vielle-Petit, L.; Wang, L.; Mallants, D.; Leterme, B. *Additional Sorption Parameters for the Cementitious Barriers of a Near-surface Repository*; ONDRAF/NIRAS: Brussels, Belgium, 2010.
17. Krishnaswamy, P. Kinetics of the Aqueous Oxidation of Molybdenite and the Role of Crystal Anisotropy on the Electrochemical Mechanisms of the Process. Ph.D. Dissertation, University of California, Berkeley, CA, USA, 1981; pp. 11–12.

minerals

MDPI

Article

Bubble Size Characterization in the HydroFloat® Fluidized-Bed Flotation Cell Using Tap Water and Seawater

Giovanni Gahona [1], Luís A. Cisternas [1,*], Natalia Araya-Gómez [2], Freddy A. Lucay [3], Edelmira D. Gálvez [4], Alejandro Lopéz-Valdivieso [5] and Felipe Valdes [6]

1 Departamento de Ingeniería Química y Procesos de Minerales, Universidad de Antofagasta, Antofagasta 1240000, Chile; giox.190@gmail.com
2 Department of Chemical and Metallurgical Engineering, School of Chemical Engineering, Aalto University, 02150 Espoo, Finland; natalia.a.arayagomez@aalto.fi
3 Escuela de Ingeniería Química, Pontificia Universidad Católica de Valparaíso, Valparaíso 2374631, Chile; freddy.lucay@pucv.cl
4 Departamento de Ingeniería Metalúrgica y Minas, Universidad Católica del Norte, Antofagasta 1240000, Chile; egalvez@ucn.cl
5 Instituto de Metalurgia, Universidad Autónoma de San Luis de Potosi, San Luis Potosi 78000, Mexico; alopez@uaslp.mx
6 Eriez Flotation Group Chile S.A., Santiago 7560908, Chile; fvaldes@eriez.com
* Correspondence: luis.cisternas@uantof.cl

Abstract: This research aims to analyze the behavior of bubble size distribution in the HydroFloat® with seawater and tap water. The study characterized bubble size in a two-phase gas–water system in a fluidized-bed flotation cell. The impact of seawater was compared to tap water using two frothers, MIBC and polyglycol F507. The experimental design was used to investigate the influence of various parameters such as superficial air velocity, superficial liquid velocity, frother concentration, and seawater concentration on bubble size. The results indicate that the critical coalescence concentration followed the order of MIBC > F507. Bubble size decreases with increasing superficial liquid velocity, while the superficial gas velocity and frother/seawater concentration have the opposite effect. ANOVA results reveal that all linear factors are significant, the quadratic terms of the frother and seawater concentrations are significant, and the interaction term for the superficial air velocity–superficial liquid velocity is nonsignificant for bubble size. Global sensitivity analysis demonstrates that the variables significantly affecting bubble size are frother concentration and seawater concentration, followed by superficial water velocity. The superficial gas velocity has minimal impact on bubble size under the conditions studied.

Keywords: flotation; bubble size; seawater; critical coalescence concentration; HydroFloat®

1. Introduction

Mineral froth flotation is widely used in the mining industry worldwide to concentrate millions of tons of ore each year due to its versatility and cost-effectiveness [1]. This process utilizes the surface property of hydrophobicity to separate fine mineral particles [1,2]. Prior to flotation, ores must be reduced to a specified top size, typically between 75 and 150 (d_{90}) [3,4]. Froth flotation is effective within a range of diameters between 20 and 150 μm for base metal ores [4]. The recovery rate decreases rapidly outside of this range. Froth flotation is primarily used to process fine minerals such as chalcopyrite, pyrite, and molybdenite [5–9].

In recent studies, froth flotation was also used for processing coarse minerals such as sphalerite, galena, and pentlandite [4,10]. One of the main challenges in froth flotation has been the concentration of coarse particles [4,10–16]. The decrease in the recovery rates of coarse sulfide particles can be explained using the "elephant curve" of flotation data developed by Lynch et al. [14]. The drop-off in recovery rates for coarse particles is related to high turbulence in conventional flotation cells, allowing the detachment of coarse particles from the bubbles [3,4,11,14,15]. The low collection efficiency for coarse particles has been linked to high detachment efficiency and low attachment efficiency [17]. Since comminution equipment is the most energy-consuming unit in mineral processing plants, the effective separation of coarse particles has become increasingly important in the mineral industry [18].

In the process of froth flotation, the use of frothers is essential [19–22]. Frothers are types of surface-active compounds or surfactants that consist of a hydrophobic part (non-polar hydrocarbon chain) and a hydrophilic part (polar chain) in their structure [8,20]. The role of frothers is to create a stable froth phase and facilitate the formation of small bubbles [8,19,21,23,24]. Cho and Laskowski [23] conducted a study on the impact of frothers on foam stability and bubble size. They found that bubble size is mainly determined by the concentration of bubble coalescence below a specific level of frother concentration known as the critical coalescence concentration (CCC). Since then, some studies have determined the CCC for different surfactants and concentration conditions [25,26]. Szyszka [26] determined the CCC of the selected ingredients in industrial flotation frothers by analyzing air bubble sizes. Nassif et al. [25] developed a method, known as the dilution method, to calculate the CCC using a commercial frother on a laboratory scale. The authors expressed the CCC as equivalent to the frother concentration to compare water samples using the plot of Sauter mean bubble size (d_{32}) against frother concentration, which is a method to characterize bubble size reduction and allows the determination of CCC [25].

Certain inorganic electrolytes, such as NaCl, KCl, $MgCl_2$, $CaCl_2$, and Na_2SO_4, have been shown to mimic the role of frothers [27]. Indeed, frothers are more effective in saline water because the inorganic electrolytes inhibit the bubbles' coalescence and stabilize the froth layer [28]. Sovechles and Waters [29] determined the CCC for inorganic salts and a synthetic sea salt solution usually present in froth flotation systems. CCC values varied from 0.02 M ($Al_2(SO_4)_3$) to 0.25 M (KCl), and the CCC decreased for multivalent ion salts. The results showed that the rupture of the multicomponent sea salt into parts featured a good correlation between the addition of the ionic strength of each ion and the overall ionic strength curve for all of the salts studied.

There is a growing interest in using seawater in mining processes due to water scarcity in places where mining usually occurs. One example is Chile, which hosts several mineral processing plants in the Atacama Desert [30]. Seawater contains several elements, mainly in the form of ions. Most (86%) of these ions comprise sodium and chloride, which results in salinity. The salinity is usually approximately 35‰, but it can reach 45‰ depending on the geographical location. Other secondary ions are magnesium, calcium, potassium, sulfate, and carbonate [27,30–32]. Salinity affects the properties of water, such as the density, viscosity, vapor pressure, and surface tension [30]. The pH of seawater varies between 7.8 and 8.2 according to the concentration of carbonate/bicarbonate ions and boric acid/borate ions [27,30].

In Chile, many copper mineral processing plants are currently using or exploring the option of using seawater directly in their operations [27]. It has been found that saline solutions can be used to replace frothers [20,33,34]. However, a critical consideration when using seawater in froth flotation is the precipitation of seawater at alkaline pH levels [32]. Some elements in seawater can interact with minerals and chemical reagents, leading to reduced recovery rates and damage to the flotation equipment. It is important to note that seawater has different physicochemical properties compared to the water typically used in mineral processing, and this difference must be taken into account [30]. Seawater contains small quantities of ions such as Mg^{2+}, Ca^{2+}, HCO_3^-, and CO_3^{2-} that can affect surface phenomena [32]. Secondary ions such as magnesium (Mg^{2+}) and calcium (Ca^{2+}) are detrimental to the flotation of molybdenum and copper [35]. Recently, a review by Cruz et al. [27] discussed the challenges of utilizing seawater in copper flotation. The review highlighted several research opportunities, including determining the impact of seawater ions on chemical reagents, evaluating the effect of seawater on emerging technologies, conducting a detailed analysis of the interactions between metal and seawater, as well as reagent and seawater, and assessing the current industrial-scale use of seawater.

Many researchers have investigated the use of seawater in froth flotation [27,30,35–37]. Jeldres et al. [31] conducted a study on copper–molybdenum sulfide ore flotation using seawater devoid of calcium and magnesium. They discovered that seawater hardness decreases, high recoveries of copper and molybdenum are achieved in highly alkaline conditions, and significant pyrite depression is attained at pH 11.5. Cruz et al. [35] suggested a method involving carbon dioxide gas and sodium hydroxide to eliminate calcium and magnesium from seawater for its use in the froth flotation of copper–molybdenum minerals, aiming to enhance recovery rates. This method removed 60.5% of calcium and 98.3% of molybdenum species, allowing the recovery of 81.1% of molybdenum and 93.4% of copper. Quinn et al. [33] compared the effect of a frother and methyl isobutyl carbinol (MIBC) with the water used in the Ranglan concentrator (Xtrata Nickel). This process does not use a frother and instead uses water with a high salinity (ca. 30,000 ppm). The objective was to compare the frother and the water performance on gas dispersion. The tests were performed in laboratory columns of two phases (solution–air) and three phases (slurry–air). The results showed that the froth texture, bubble size distribution, and froth overflow rate obtained with salt solutions are similar to those obtained using the usual frother dosage.

Bubble size is affected by the break-up phenomena and bubble coalescence, and has a crucial role in the kinetics of froth flotation [29]. Both frothers and seawater can stabilize and reduce bubble size [29]. Arancibia-Bravo et al. [36] studied the influence of various parameters on d_{32} for saline solutions on a laboratory scale using the response surface methodology and ANOVA. The results showed that the salt concentration under the conditions studied was the only one to affect all of the saline solutions studied. Zhu et al. [38] studied bubble size evolution in a Jameson cell. The authors investigated the influence of three parameters on bubble size: MIBC concentration, superficial liquid velocity (J_l), and superficial air velocity (J_g). This study showed that J_l and J_g have the opposite effect on the Reynolds (Re) number and bubble size. An increase in J_l increases the Re number because the solution's density, viscosity, and velocity increase. The opposite happens in terms of the increase in J_g because the velocity increases, but the density and viscosity of the solution decrease, causing a reduction in the Re number.

Fluidized beds such as the Hydrofloat® cell have been used recently to improve the flotation of coarse particles [11,14,15,39,40]. The HydroFloat® cell was designed in early 2000. It was the first embodiment to approach the drop-off in the recovery of sulfide particles within the mineral flotation plant; where particles > 150 μm, the recovery decays considerably, and this has been previously illustrated by the size-by-size flotation data in the well-recognized "elephant curve" [14,41]. The Hydrofloat® offers several advantages over traditional flotation cells. It minimizes turbulence by creating a calm environment inside the cell, which allows coarse particles to attach to the bubbles. This results in longer retention time, reduced particle buoyancy constraints, and minimal detachment of coarse particles from the bubbles. Additionally, it has an increased capacity for bubble–liquid segregation during the process, which leads to excellent coarse particle recovery (references [14,15,18,39]). Despite the numerous scientific publications on the bubble size in conventional flotation cells using frother or ionic salts, there have been no studies investigating the HydroFloat® cell. This study aims to comprehend the behavior of bubble size distribution in the Hydrofloat® cell using seawater and tap water. Moreover, this study introduces the use of seawater in the HydroFloat® cell for the first time.

2. Methodology

In this study, we conducted tests on two phases (water and air) in the HydroFloat® cell using tap water and various seawater concentrations. We examined different seawater concentrations to understand their effect on bubble size better, especially considering the use of diluted seawater as a potential way to mitigate the impact of magnesium and calcium ions. Initially, we collected samples from nine positions to assess the distribution of bubbles.

Samples were taken at vertical positions every 10 cm and horizontal positions every 2 cm, starting from the cell's center point.

MIBC with a concentration of 7.5 ppm was used to determine the CCC and the effect of both frothers on bubble size. To study the impact of the frother, seawater, tap water, J_g, and J_l on bubble size, design of experiments (DOE) was used, including three variables and three levels.

In this study, J_g and J_l are referred to as volumetric gas flow rates and volumetric liquid flow rates per unit cross-section of the Hydrofloat® flotation cell, respectively [38,42].

This section is divided into three subsections to provide a further understanding of this study:

- Apparatus and material reagents;
- Procedures;
- Mathematical modeling.

2.1. *Apparatus and Material Reagents*

The HydroFloat cell® used for this study consisted of an acrylic cylinder with a diameter of 15 cm and a height of 37 cm. The cell volume was 9 L, and the bottom had a conical shape with a height of approximately 10 cm. Figure 1 shows the schematic diagram of the system consisting mainly of the Hydrofloat® cell. As is shown in Figure 1, the main components were (1) a feed open tank for tap water mixed with the frother and receiving the overflow of the flotation cell; (2) a centrifugal pump of 1 HP for pressurizing aqueous solutions into the cell; (3) flowmeters for measuring and controlling the water and air flow rate; (4) an ejector that mixed water and air for pumping the mixture to the cell, which has 12 nozzles of 1/8″ diameter; (5) an acrylic tube with a total height of 90 cm, which collected the generated bubbles; and (6) a photo capture system to measure the bubble size.

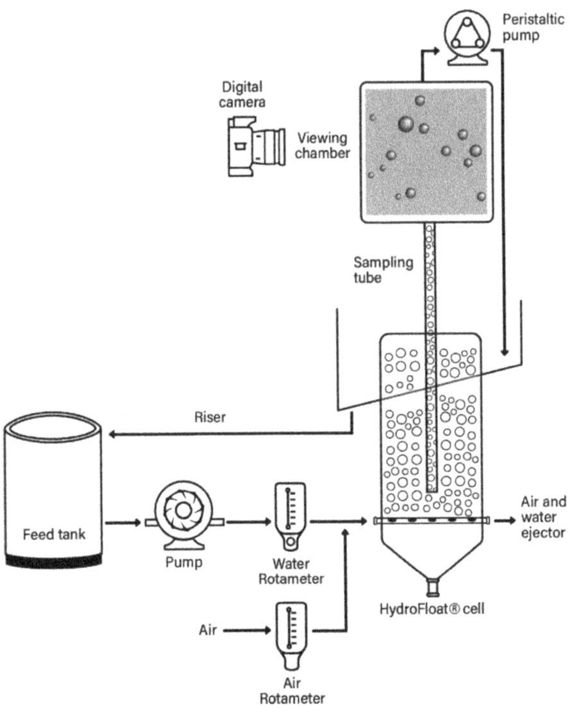

Figure 1. Schematic diagram of the system for measuring the size of bubbles in the HydroFloat® cell.

The bubble measurement system used a Masterflex L/S peristaltic pump to sample bubbles from the solution 3 cm above the ejector at a flow rate of 2.1 L/min. The system operated continuously, and any overflow from the column was recycled back to the feed tank to ensure solution homogeneity. All tests were conducted after the system had been running for 5 min. The procedure for measuring bubble size had been validated in previous research [36], which compared our results to those obtained by Finch et al. [20]. Experimental deviations in bubble diameter were found to be within ±0.02 mm.

The photo capture system included a Nikon D610 DSLR camera with an AF-S Micro Nikon 105 mm 1:2.8 G lens. The typical image resolution was 60 pixels/mm and the camera took one frame per second. The area for collecting images was 42 mm × 34 mm, which was controlled using Nikon Software Camera Control Pro 2 v2.28.0. To determine the distribution of bubble sizes, Fiji software via ImageJ v1.52b was utilized. A minimum of 5000 bubbles were measured for each tested condition.

Two frothers, AEROFROTH 70 and Oreprep F507, provided by Solvay with a purity of over 90%, were used in this study for tap water experiments. AEROFROTH 70 is a commercial name for MIBC, an alcohol frother of the molecular formula $(CH_3)_2C_3H_4OHCH_3$ and a molecular weight of 102.2 g/mol [24]. Oreprep F507 is a polyglycol frother with a molecular weight of 425 g/mol. Bubble size measurements were conducted at the ambient pH of tap water, which was measured at pH 7.2.

The seawater tests used seawater from San Jorge Bay, located in Antofagasta, northern Chile. Table 1 shows the chemical composition of the seawater sample [43].

Table 1. Seawater composition at San Jorge Bay in Antofagasta (adapted from Arias et al. [43]).

Parameter	San Jorge Bay Seawater Concentration (mg/L)
Magnesium (Mg^{2+})	1310 ± 38
Sodium (Na^+)	11,138 ± 12
Potassium (K^+)	401 ± 4
Calcium (Ca^{2+})	415 ± 26
Chloride (Cl^-)	19,867 ± 24
Nitrate (NO_3^-)	3.62 ± 0.38
Bicarbonate (HCO_3^-)	143 ± 5
Sulfate (SO_4^{2-})	2791 ± 18

2.2. Procedures

The data generated from Fiji were automatically transferred to Excel and analyzed via Visual Basic for Applications (VBA) to determine the bubble size distribution (BSD), and Sauter mean diameter (d_{32}), which was calculated using the expression that determines the diameter based on the bubble area, as shown in Equations (1) and (2) [44]:

$$d_a = \sqrt{\frac{4A_p}{\pi}} \tag{1}$$

$$d_{32} = \frac{\sum_{i=1}^{n} d_i^3}{\sum_{i=1}^{n} d_i^2} \tag{2}$$

where d_a is the diameter of the bubble particle, A_p is the area of the bubble particle given by Fiji, d_i is the diameter of the i bubble, and n is the number of the overall sample.

Nesset et al. [44] suggested fitting the Sauter mean diameter vs. frother concentration following the three-parameter model:

$$d_{32} = d_L + A\ Exp[-B \cdot C] \tag{3}$$

where d_L is the limit that d_{32} can reach as the frother concentration tends to be infinite, A is the difference between d_L and the initial d_{32} in water (without the frother), B is the decay constant, and C is the frother concentration. This model calculates the CCC_X, which is the concentration where d_{32} is reduced by X %.

$$CCC_X = -\frac{In(1-x)}{B} \quad (4)$$

2.3. Mathematical Modeling

A full factorial design [45] was used to design the experiments to identify the input factors that affect the bubble size, and replicas of the central point were used to obtain an independent estimation of the experimental error (Table 2). Then, response surface methodology was used to model the data obtained [46]. The ANOVA test and global sensitivity analysis, using the Sobol–Jansen method [47], were utilized to analyze the importance of the factors. The Sobol–Jansen method was selected because it performs best among several Sobol methods [48] and has been successfully used in other mineral processing studies [49,50]. Interpretation of the Sobol–Jansen indices is straightforward: the higher its value for a factor, the greater the effect of that factor on the output or dependent variable.

Table 2. Factors and their levels for full factorial experiments for MIBC and seawater.

Type of Experiment	Factor Bubble Size Measure	Coded Variable Level		
		Low −1	Center 0	High 1
Experiment A	X_g: Superficial air velocity, cm/s	1.35	2.70	4.06
	X_l: Superficial water velocity, cm/s	13.5	17.6	21.7
	X_c: Frother concentration, ppm	5.0	10.0	15.0
Experiment B	X_g: Superficial air velocity, cm/s	1.35	2.70	4.06
	X_l: Superficial water velocity, cm/s	13.5	17.6	21.7
	X_S: Seawater in solution, mol/L	0.053	0.291	0.529

3. Results and Discussion

3.1. Effect of Different Position Samples on Bubble Size in the HydroFloat® Cell

Figure 2 shows the d_{32} at nine different positions in the HydroFloat® cell using 7.5 ppm MIBC in tap water. The bubble size ranges from 0.96 to 1.04 mm, with an average of 1.00 mm ± 0.035 mm and a standard deviation of approximately 3.6% in the different areas.

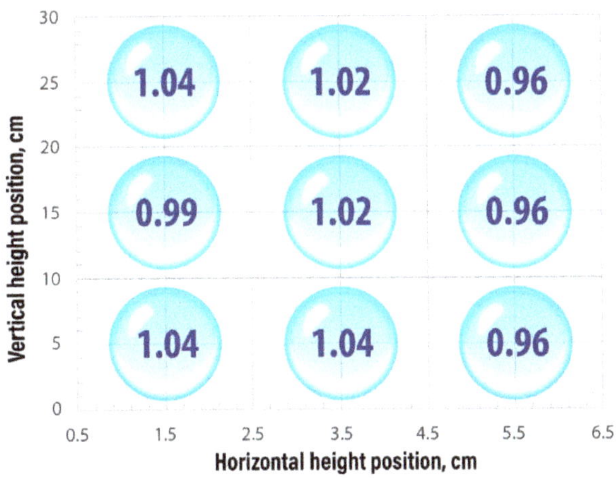

Figure 2. Sauter mean diameter in different positions on HydroFloat® flotation cell using 7.5 ppm MIBC at J_g = 2.70 cm/s and J_l = 17.6 cm/s.

The results in Figure 3 confirm the findings in Figure 2 regarding the average bubble size distribution (BSD) for both vertical and horizontal positions. The bubble size remains consistent across all positions and reaches a maximum size of 1.00 mm, accounting for an average of 33% ± 2.71% of the total bubbles. These consistent results suggest that the cell is representative, likely due to its quiescent state. The absence of turbulence zones, characteristic of conventional flotation cells with mechanical stirrers, ensures that the bubble size remains unaffected. Therefore, subsequent tests will be conducted at the central bottom position, 3 cm above the ejector, at a vertical height of 5 cm and a horizontal height of 3.5 cm. It is important to note that these findings may not apply to different operational conditions and types of frothers, but they do support reasonable assumptions.

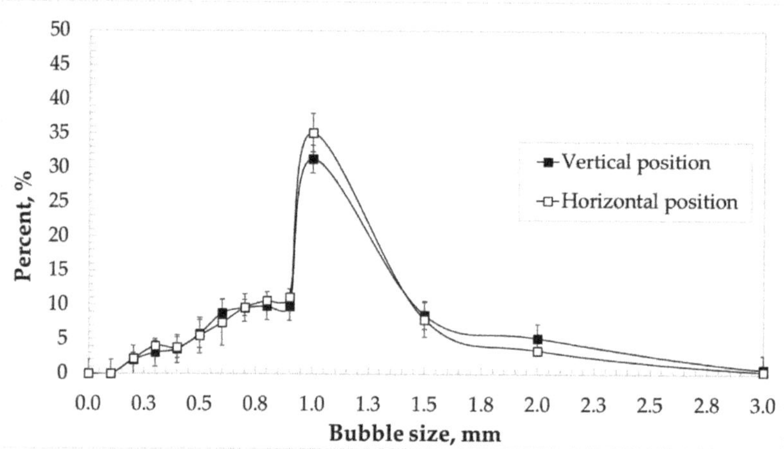

Figure 3. Bubble size distribution average using 7.5 ppm MIBC in a vertical and horizontal position at J_g = 2.70 cm/s and J_l = 17.6 cm/s.

3.2. *Effect of Frother Reagents on Bubble Size Using Tap Water*

Figure 4 compares MIBC and F507 frother reagents in d_{32} as a function of the frother concentration. The d_{32} value decreased with the increased frother concentration until it reached the CCC, which maintained the bubble size constant at 0.55 and 0.46 mm for MIBC and F507, respectively. Where the reagent concentration is low, the bubble size is larger, and bubble coalescence may occur. However, in areas where the concentration is higher, the bubble size drops significantly to an area where the bubble size is not concentration dependent, resulting in stable bubbles. This transition concentration for bubble coalescence stability has been explained by several authors [23,51,52].

Equation (3) was used to fit the experimental data for both reagents, obtaining the following equations:

$$d_{32} \text{ of MIBC} = 0.510 + 1.64\, Exp[-25 \cdot C] \quad (5)$$

$$d_{32} \text{ of F507} = 0.400 + 1.75\, Exp[-55 \cdot C] \quad (6)$$

Using Equation (1), we determined the CCC values for each frother at J_g = 2.70 cm/s and J_l = 17.6 cm/s. For MIBC, the CCC was 11.98 ppm, equivalent to 0.1175 mmol/L; for F507, the CCC was 5.47 ppm or 0.0128 mmol/L. The CCC value was lower for higher molecular weights; MIBC has a molecular weight of 102 g/mol, which is lower than that of F507 at 425 g/mol. This trend was similar in other studies [21,23,51]. Cho and Laskowski [23] determined the CCC for five frothers, including MIBC and four different hexanol isomers, finding that bubble size is a product of coalescence at frother concentrations lower than the

CCC. Corona-Arroyo et al. [51] determined the CCC for three frothers, MIBC, DDA, and F507, in a downflow column, finding the same trend of lower values of the CCC for higher molecular weights. The same trend was found previously by Melo and Laskowski [21] by testing three frothers, MIBC, DF-200, and DF-1012; the CCC was lower for higher molecular weights.

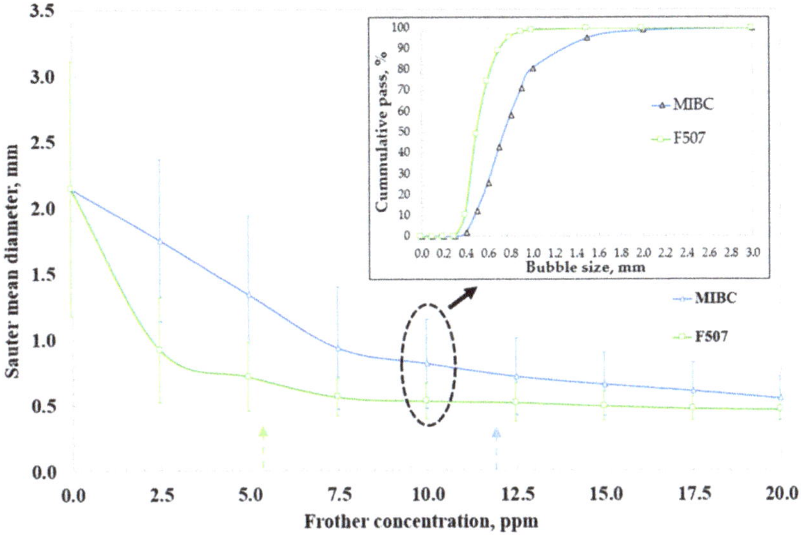

Figure 4. Sauter mean diameter as a function of frother concentration in HydroFloat® system for J_g = 2.70 cm/s and J_l = 17.6 cm/s. The insert is the bubble size distribution at a concentration of MIBC and F507 at 10 ppm.

The minimum sizes of the bubbles produced using MIBC and F507 were found to have an inverse correlation with their CCC values. It has been observed that frothers impede coalescence and influence break-up [20]. According to the proposed break-up mechanism, frothers cause surface tension gradients, which increase instabilities along the air/water interface. Compared to F507, the MIBC molecule is smaller. It has fewer hydrophilic sites for H-bonding with water molecules, which means that the MIBC molecules on the bubble surface are closer, resulting in smaller break-away bubbles.

The arrow lines in Figure 4 indicate these CCC values. In the HydroFloat® flotation cell, the bubble size can achieve lower d_{32} values than 0.630 mm, on average, in conventional flotation cells [20,25,37,38]. In Figure 4, the insert shows that F507 produces a finer BSD than MIBC due to its longer hydrocarbon chain and strong and active surface [51].

Figure 5 shows the bubble size distribution of five different concentrations for MIBC and F507. The distribution follows other results in the literature [23,51]. Cho and Laskowski [23] studied bubble size distribution and foam stability using different frothers, finding that frothers control bubble size; at lower concentrations than the CCC, coalescence determines bubble size.

As the frother concentration increases, the bubble size tends to decrease, resulting in a solution dominated mainly by a bubble size of <0.6 mm. As the frother concentration increases, smaller and more stable bubbles are produced. On the other hand, a low concentration of F507 produces more small and stable bubbles than MIBC. In the Hydrofloat® cell, a more significant number of bubbles smaller than 0.45 mm is obtained using a concentration greater than the CCC.

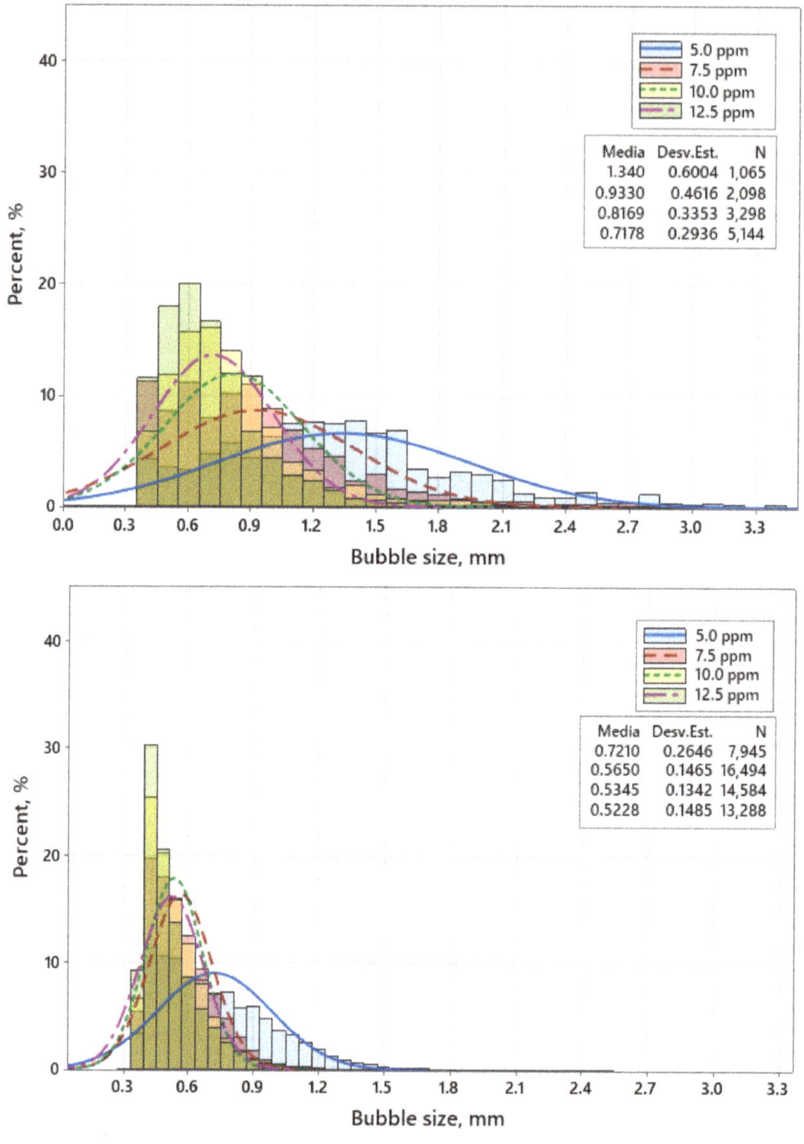

Figure 5. Bubble size distributions at different frother concentrations from 5.0 to 12.5 ppm for (**top**) MIBC and (**bottom**) F507 at J_g = 2.70 cm/s and J_l = 17.6 cm/s.

3.3. *Effect of J_g and J_l on Bubble Size Using Tap Water*

Figure 6a shows the d_{32} value as a function of J_g for MIBC. When the J_g increases, the d_{32} increases at all concentrations; this variation becomes more noticeable at lower frother concentrations because the adsorption on the bubble surface is lower.

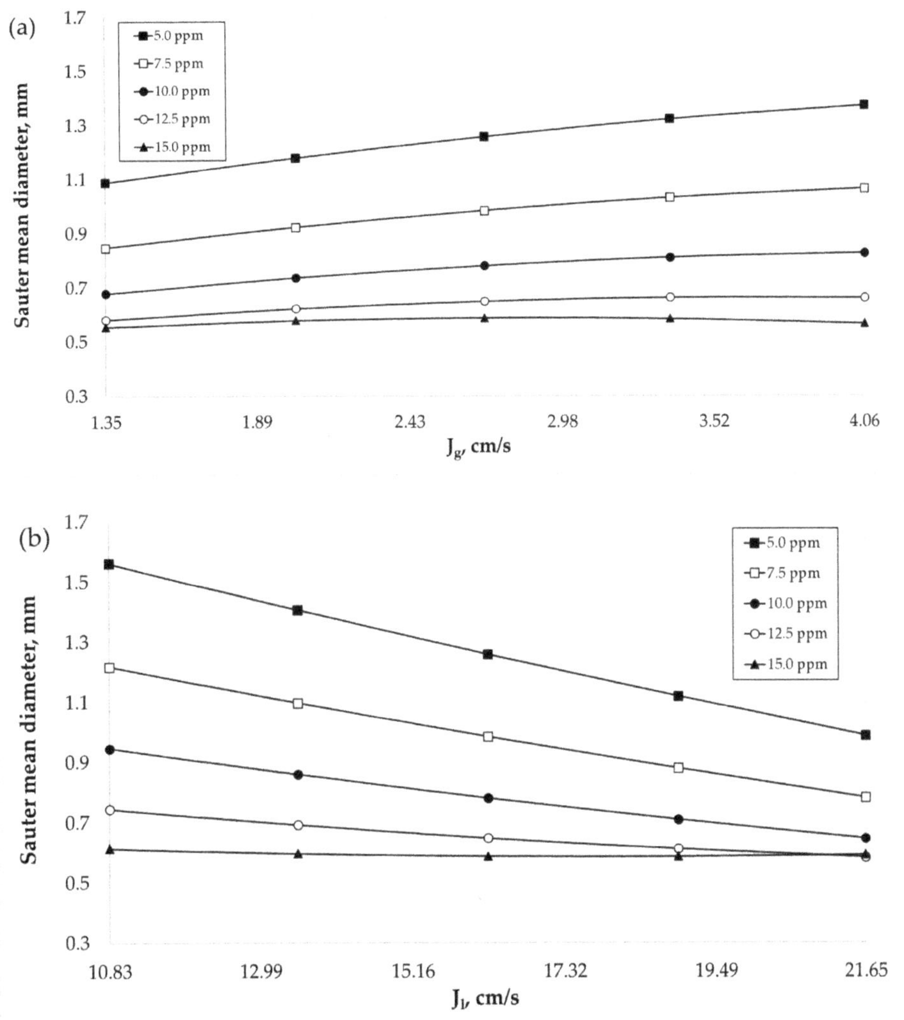

Figure 6. Sauter mean diameter as a function of J_g and J_l for different concentrations of MIBC. (**a**) Function of J_g for J_l = 17.6 cm/s and (**b**) function of J_l for J_g = 2.70 cm/s.

This increase in bubble size is related to the mass of air injected so that the variations increase and provide more air mass to the system, causing larger bubbles. When the concentration exceeds the CCC, the variation is not appreciated. The bubble size is limited to stabilize its minimum size; in low concentrations, the bubble size difference can reach 300–250 μm, and in high concentrations, the difference can be 80 μm [20].

Figure 6b shows the variation in d_{32} as a function of J_l. When J_l increases, d_{32} decreases at all concentrations. This variation was opposite to the effect of J_g, where a lower frother concentration led to a greater reduction in d_{32}. At low concentrations, the bubble size difference can reach 600–500 μm, and in high concentrations, the difference can be 250 μm. The adjustment of J_g and J_l causes the Re number variation in the HydroFloat® cell. Increasing J_g leads to an increase in the velocity of the solution but a decrease in the density and viscosity of the solution, inducing the reduction in the Re number. We can see the opposite effect when J_l increases, increasing density and viscosity and increasing Re [38]. The impact

The coefficient of determination, R squared, was 97.0% and 97.5% for MIBC and seawater, respectively, meaning that 97.0% and 97.5% of the variance in the bubble size can be explained by the model of Equations (7) and (8), respectively. The R squared adjusted was 95.9% and 96.6% for MIBC and seawater, respectively. Also, the root-mean-square error (RMSE) was 0.05 and 0.07 mm, respectively. Figure 9 shows the response surface when two variables change up or down while the other stays at a central point.

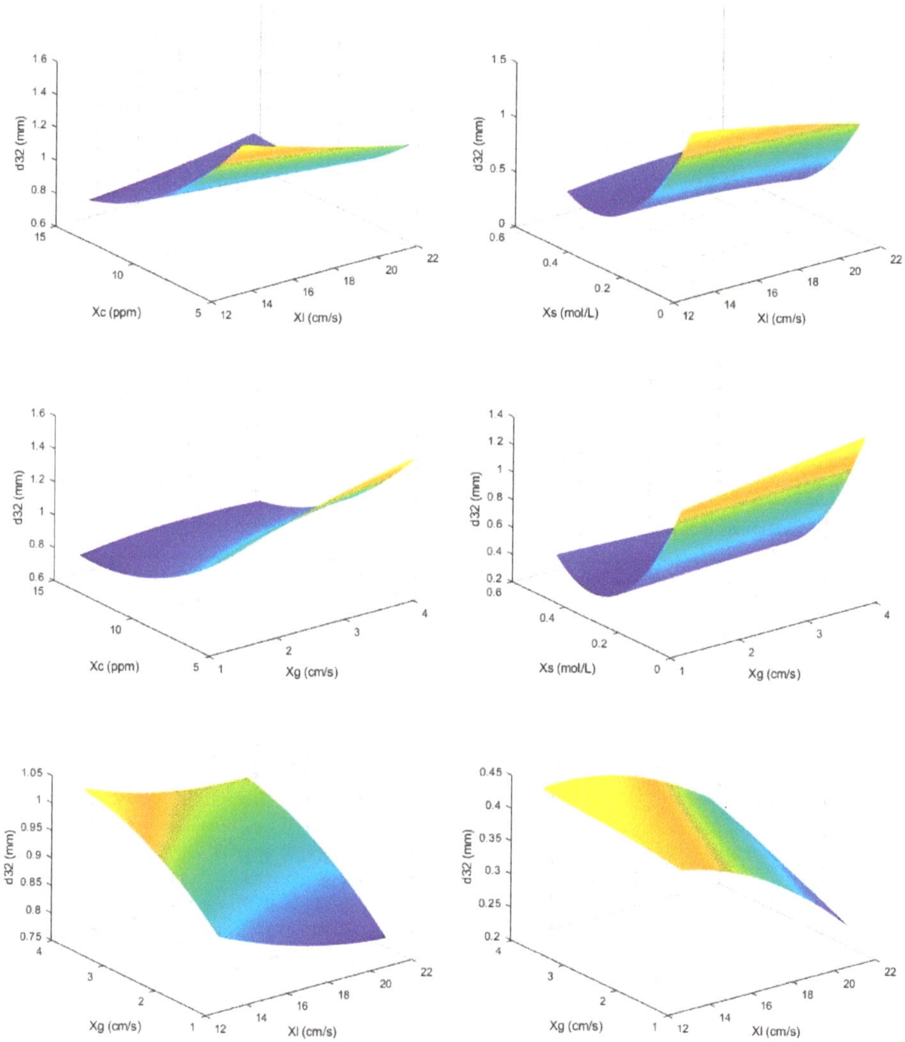

Figure 9. Contour surface as a function of X_g, X_l, X_c, and X_s on the bubble size.

The analysis of variance, ANOVA, test was used to identify the terms of Equations (7) and (8) that are significant in predicting the bubble size. Table 4 shows the terms under study, where the significant terms present a p-value smaller than 0.05 (significant terms are identified with the symbol ✓; non-significant terms are identified with the symbol ×). The global sensitivity analysis results based on the Sobol–Jansen method are also included in Table 4. The variables that significantly affect the system are frother concentration at 0.718

and seawater concentration at 0.948; the superficial water velocity has a secondary role with values of 0.167 and 0.013 in MIBC and seawater, respectively. These results clearly show us that the solute concentration in the system has the most significant effect on the flotation cells. Interactions of the combination variables also show that they tend to decrease the bubble size, although some variables have a more substantial impact.

Table 4. Identification of influential factors on bubble size (significant ✓; non-significant ×).

Tap Water Experiments Using MIBC in Frother Experiments			Seawater Experiments without Using a Frother		
Term Constant	**Result**	**Sobol–Jansen Total Index**	**Term Constant**	**Result**	**Sobol–Jansen Total Index**
X_g	✓	0.010	X_g	✓	0.006
X_l	✓	0.167	X_l	✓	0.013
X_c	✓	0.718	X_s	✓	0.948
$X_g \times X_l$	×	0.222	$X_g \times X_l$	×	0.022
$X_g \times X_c$	✓	0.763	$X_g \times X_s$	✓	0.966
$X_l \times X_c$	✓	0.952	$X_l \times X_s$	✓	0.981
${X_g}^2$	×	-	${X_g}^2$	×	-
${X_l}^2$	×	-	${X_l}^2$	×	-
${X_c}^2$	✓	-	${X_s}^2$	✓	-
R-Sq	97.1%		R-Sq	97.7%	
R-Sq(adj)	95.5%		R-Sq(adj)	96.5%	

Factors such as water or air flow do not affect bubble size over a given seawater concentration [34,36]. Airflow is directly proportional to the size of the bubble, while water flow is inversely proportional. Other studies have demonstrated these results [24,36,44,51].

As shown in Table 4, in the tests with MIBC, the variables affecting the process the most are X_c at 0.71, X_l at 0.16, and X_g at 0.01%. For seawater, X_s (mainly NaCl) affects the size of the bubble by 0.948. For both experiments, the solute has the most significant impact on the change in bubble size, but in seawater, the variables X_g and X_l tend to suppress their effect. It is known that airflow and water flow affect the size of bubbles [34,44], but the effect of seawater concentration can eliminate those factors and generate stable bubbles without changing their size.

These results can also be analyzed using Figure 6; increasing the airflow, the bubble size tends to increase, while increasing the water flow, the bubble size tends to decrease. Both factors can be expected to be eliminated when seawater exceeds 50% of the concentration in the solution.

4. Conclusions

The study yields the following conclusions:

- The tap water experimental results indicate that the Sauter mean bubble size (d_{32}) decreased as the concentration of frother increased, approaching the critical coalescence concentration (CCC). At gas superficial velocity (J_g) = 2.70 cm/s and liquid superficial velocity (J_l) = 17.6 cm/s, the CCC for MIBC was determined to be 11.98 ppm, corresponding to 0.1175 mmol/L, while for F507, the CCC was found to be 5.47 ppm, which is equivalent to 0.0128 mmol/L.
- A higher concentration of the frother produces smaller stable bubbles. Comparing F507 with MIBC, a lower concentration of the first produces more diminutive and more stable bubbles, with a constant size of 0.51 and 0.40 mm for MIBC and F507, respectively. This variation in bubble size can be attributed to frothers that impede coalescence and also influence break-up. In comparison to F507, the MIBC molecule is smaller, possessing fewer hydrophilic sites for bonding with water molecules. Consequently, the MIBC molecules on the bubble surface are closer, resulting in smaller break-away bubbles.

- The analysis of the effect of J_g and J_l on bubble size using an MIBC frother in experiments with tap water showed that when J_g increases, d_{23} increases for all frother concentrations. This is more noticeable with a lower concentration of MIBC. The opposite situation occurs when J_l increases.
- In the experiments with different seawater concentrations, when the seawater concentration increases, bubble size decreases, reaching a minimum size of 0.41 mm. Comparing the results of seawater with tap water using a frother, we can observe the positive effects of seawater on the bubble's compression, even when a lower concentration of seawater is used. With 40% seawater in the solution, the bubble size begins to stabilize, and 87% of the bubbles reach sizes between 0.28 and 0.4 mm. The CCC was obtained with 51% seawater. Hence, we can use 40 and 50% seawater solutions to achieve a minimum bubble size.
- The design of experiment technique was used to examine how J_l and J_g, frother concentration, and seawater concentration affect bubble size. The results indicate that all linear factors are significant, the quadratic terms of the frother and seawater concentrations are also significant, and the interaction term between superficial air velocity and superficial liquid velocity does not significantly impact bubble size. Through global sensitivity analysis, it is observed that the variables with the most significant impact on bubble size are the frother and seawater concentrations, followed by superficial liquid velocity. Superficial gas velocity has little effect on bubble size under the conditions studied.

Saline solutions have been proposed to replace frothers in froth flotation. This study shows that seawater has a greater impact on bubble sizes compared to the frother, resulting in smaller bubble sizes and a narrower distribution. Using seawater in froth flotation in areas of water scarcity would help reduce the stress on continental water resources. It would also decrease water treatment costs for mining companies.

The flotation process is a complex separation process influenced by numerous factors. The size of the bubbles is affected by various parameters such as the frother, collector, particle and liquid properties, hydrodynamics, energy input, and airflow rate within the flotation cell. Subsequent studies should thoroughly evaluate the combined effects of these factors on the Hydrofloat®.

Author Contributions: Conceptualization, G.G., L.A.C., A.L.-V. and E.D.G.; Methodology, G.G., L.A.C. and F.A.L.; Validation, G.G. and N.A.-G.; Investigation, G.G. and L.A.C.; Writing—Original Draft, G.G. and N.A.-G.; Writing—Review and Editing, L.A.C.; Supervision, L.A.C., E.D.G., A.L.-V. and F.V.; Formal Analysis, E.D.G. and F.A.L.; Funding Acquisition, L.A.C. and E.D.G. All authors have read and agreed to the published version of the manuscript.

Funding: This research received no external funding.

Data Availability Statement: All data underlying the results are available as part of the article and no additional source data are required.

Acknowledgments: The authors are grateful for the support of Agencia Nacional de Investigación y Desarrollo de Chile (ANID) through grants ANID/ACT210027, MEC80190019, and Fondecyt 1240182. N.A.-G. would also like to thank the Finnish Foundation for Technology Promotion (Tekniikan Edistämissäätiö) for grant 8015.

Conflicts of Interest: Felipe Valdes is an employee of Eriez Manufacturing Company. The authors declare that they have no known competing financial interests or personal relationships that could have appeared to influence the work reported in this paper.

References

1. Mesa, D.; Brito-Parada, P.R. Scale-up in froth flotation: A state-of-the-art review. *Sep. Purif. Technol.* **2019**, *210*, 950–962. [CrossRef]
2. Lai, H.; Deng, J.; Liu, Q.; Wen, S.; Song, Q. Surface chemistry investigation of froth flotation products of lead-zinc sulfide ore using ToF-SIMS and multivariate analysis. *Sep. Purif. Technol.* **2021**, *254*, 117655. [CrossRef]
3. Jameson, G.J. New directions in flotation machine design. *Miner. Eng.* **2010**, *23*, 835–841. [CrossRef]
4. Jameson, G.J. Advances in fine and coarse particle flotation. *Can. Metall. Q.* **2010**, *49*, 328–330. [CrossRef]

5. Cilek, E.C.; Ozgen, S. Improvement of the flotation selectivity in a mechanical flotation cell by ultrasound. *Sep. Sci. Technol.* **2010**, *45*, 572–579. [CrossRef]
6. Hirajima, T.; Mori, M.; Ichikawa, O.; Sasaki, K.; Miki, H.; Farahat, M.; Sawada, M. Selective flotation of chalcopyrite and molybdenite with plasma pre-treatment. *Miner. Eng.* **2014**, *66*, 102–111. [CrossRef]
7. Hirajima, T.; Miki, H.; Suyantara, G.P.W.; Matsuoka, H.; Elmahdy, A.M.; Sasaki, K.; Imaizumi, Y.; Kuroiwa, S. Selective flotation of chalcopyrite and molybdenite with H_2O_2 oxidation. *Miner. Eng.* **2017**, *100*, 83–92. [CrossRef]
8. Khoshdast, H. Flotation Frothers: Review of Their Classifications, Properties and Preparation. *Open Miner. Process. J.* **2011**, *4*, 25–44. [CrossRef]
9. Yu, Y.; Ma, L.; Cao, M.; Liu, Q. Slime coatings in froth flotation: A review. *Miner. Eng.* **2017**, *114*, 26–36. [CrossRef]
10. Awatey, B.; Skinner, W.; Zanin, M. Effect of particle size distribution on recovery of coarse chalcopyrite and galena in Denver flotation cell. *Can. Metall. Q.* **2013**, *52*, 465–472. [CrossRef]
11. Awatey, B.; Thanasekaran, H.; Kohmuench, J.N.; Skinner, W.; Zanin, M. Optimization of operating parameters for coarse sphalerite flotation in the HydroFloat fluidised-bed separator. *Miner. Eng.* **2013**, *50–51*, 99–105. [CrossRef]
12. Mehrfert, P.J. Investigating the Potential of HydroFloat (TM) Coarse Particle Flotation Techniques on Copper Sulphide Ores. In Proceedings of the 49th Annual Canadian Minerals Processors Conference Proceedings, Ottawa, ON, Canada, 26 May 2017; ALS Metallurgy: Kamloops, BC, Canada, 2017.
13. Hassanzadeh, A.; Safari, M.; Hoang, D.H. Fine, coarse and fine-coarse particle flotation in mineral processing with a particular focus on the technological assessments. In Proceedings of the 2nd International Conference on Mineral Science, Online, 1–15 March 2021; p. 9383.
14. Kohmuench, J.N.; Mankosa, M.J.; Thanasekaran, H.; Hobert, A. Improving coarse particle flotation using the HydroFloat™ (raising the trunk of the elephant curve). *Miner. Eng.* **2018**, *121*, 137–145. [CrossRef]
15. Kohmuench, J.N.; Luttrell, G.H.; Mankosa, M.J. Coarse particle concentration using the HydroFloat separator. *Miner. Metall. Process.* **2001**, *18*, 61–67. [CrossRef]
16. Lawagon, C.P.; Nisola, G.M.; Cuevas, R.A.I.; Torrejos, R.E.C.; Kim, H.; Lee, S.-P.; Chung, W.-J. $Li_{1-x}Ni_{0.5}Mn_{1.5}O_4$/Ag for electrochemical lithium recovery from brine and its optimized performance via response surface methodology. *Sep. Purif. Technol.* **2019**, *212*, 416–426. [CrossRef]
17. Darabi, H.; Koleini, S.M.J.; Deglon, D.; Rezai, B.; Abdollahy, M. Investigation of bubble-particle attachment, detachment and collection efficiencies in a mechanical flotation cell. *Powder Technol.* **2020**, *375*, 109–123. [CrossRef]
18. Kromah, V.; Powoe, S.B.; Khosravi, R.; Neisiani, A.A.; Chelgani, S.C. Coarse particle separation by fluidized-bed flotation: A comprehensive review. *Powder Technol.* **2022**, *409*, 117831. [CrossRef]
19. Drzymala, J.; Kowalczuk, P. Classification of Flotation Frothers. *Minerals* **2018**, *8*, 53. [CrossRef]
20. Finch, J.A.; Nesset, J.E.; Acuña, C. Role of frother on bubble production and behaviour in flotation. *Miner. Eng.* **2008**, *21*, 949–957. [CrossRef]
21. Melo, F.; Laskowski, J.S. Fundamental properties of flotation frothers and their effect on flotation. *Miner. Eng.* **2006**, *19*, 766–773. [CrossRef]
22. Ding, S.; Yin, Q.; He, Q.; Feng, X.; Yang, C.; Gui, X.; Xing, Y. Role of hydrophobic fine particles in coarse particle flotation: An analysis of bubble-particle attachment and detachment. *Colloids Surfaces A Physicochem. Eng. Asp.* **2023**, *662*, 130980. [CrossRef]
23. Cho, Y.S.; Laskowski, J.S. Effect of flotation frothers on bubble size and foam stability. *Int. J. Miner. Process.* **2002**, *64*, 69–80. [CrossRef]
24. Zhou, X.; Jordens, A.; Cappuccitti, F.; Finch, J.A.; Waters, K.E. Gas dispersion properties of collector/frother blends. *Miner. Eng.* **2016**, *96–97*, 20–25. [CrossRef]
25. Nassif, M.; Finch, J.A.; Waters, K.E. Developing critical coalescence concentration curves for industrial process waters using dilution. *Miner. Eng.* **2013**, *50–51*, 64–68. [CrossRef]
26. Szyszka, D. Critical coalescence concentration (Ccc) for surfactants in aqueous solutions. *Minerals* **2018**, *8*, 431. [CrossRef]
27. Cruz, C.; Botero, Y.L.; Jeldres, R.I.; Uribe, L.; Cisternas, L.A. Current Status of the Effect of Seawater Ions on Copper Flotation: Difficulties, Opportunities, and Industrial Experience. *Miner. Process. Extr. Metall. Rev.* **2022**, *43*, 545–563. [CrossRef]
28. Suyantara, G.P.W.; Hirajima, T.; Miki, H.; Sasaki, K. Floatability of molybdenite and chalcopyrite in artificial seawater. *Miner. Eng.* **2018**, *115*, 117–130. [CrossRef]
29. Sovechles, J.M.; Waters, K.E. Effect of Ionic Strength on Bubble Coalescence in Inorganic Salt and Seawater Solutions. *AIChE J.* **2015**, *61*, 2489–2496. [CrossRef]
30. Cisternas, L.A.; Gálvez, E.D. The use of seawater in mining. *Miner. Process. Extr. Metall. Rev.* **2017**, *39*, 1–16. [CrossRef]
31. Jeldres, R.I.; Arancibia-Bravo, M.P.; Reyes, A.; Aguirre, C.E.; Cortes, L.; Cisternas, L.A. The impact of seawater with calcium and magnesium removal for the flotation of copper-molybdenum sulphide ores. *Miner. Eng.* **2017**, *109*, 10–13. [CrossRef]
32. Castro, S. Physico-chemical factors in flotation of Cu-Mo-Fe ores with seawater: A critical review. *Physicochem. Probl. Miner. Process.* **2018**, *54*, 1223–1236.
33. Quinn, J.J.; Kracht, W.; Gomez, C.O.; Gagnon, C.; Finch, J.A. Comparing the effect of salts and frother (MIBC) on gas dispersion and froth properties. *Miner. Eng.* **2007**, *20*, 1296–1302. [CrossRef]
34. Sovechles, J.M.; Lepage, M.R.; Johnson, B.; Waters, K.E. Effect of gas rate and impeller speed on bubble size in frother-electrolyte solutions. *Miner. Eng.* **2016**, *99*, 133–141. [CrossRef]

35. Cruz, C.; Reyes, A.; Jeldres, R.I.; Cisternas, L.A.; Kraslawski, A. Using Partial Desalination Treatment To Improve the Recovery of Copper and Molybdenum Minerals in the Chilean Mining Industry. *Ind. Eng. Chem. Res.* **2019**, *58*, 8915–8922. [CrossRef]
36. Arancibia-Bravo, M.P.; Lucay, F.A.; López, J.; Cisternas, L.A. Modeling the effect of air flow, impeller speed, frother dosages, and salt concentrations on the bubbles size using response surface methodology. *Miner. Eng.* **2019**, *132*, 142–148. [CrossRef]
37. Quinn, J.J.; Sovechles, J.M.; Finch, J.A.; Waters, K.E. Critical coalescence concentration of inorganic salt solutions. *Miner. Eng.* **2014**, *58*, 1–6. [CrossRef]
38. Zhu, H.; Valdivieso, A.L.; Zhu, J.; Song, S.; Min, F.; Corona Arroyo, M.A. A study of bubble size evolution in Jameson flotation cell. *Chem. Eng. Res. Des.* **2018**, *137*, 461–466. [CrossRef]
39. Fosu, S.; Awatey, B.; Skinner, W.; Zanin, M. Flotation of coarse composite particles in mechanical cell vs. the fluidised-bed separator (The HydroFloat^{TM}). *Miner. Eng.* **2015**, *77*, 137–149. [CrossRef]
40. Islam, M.T.; Nguyen, A.V. Effect of particle size and shape on liquid–solid fluidization in a HydroFloat cell. *Powder Technol.* **2021**, *379*, 560–575. [CrossRef]
41. Trahar, W.J. A rational interpretation of the role of particle size in flotation. *Int. J. Miner. Process.* **1981**, *8*, 289–327. [CrossRef]
42. Majumder, S.K.; Kundu, G.; Mukherjee, D. Mixing mechanism in a modified co-current downflow bubble column. *Chem. Eng. J.* **2005**, *112*, 45–55. [CrossRef]
43. Arias, D.; Villca, G.; Pánico, A.; Cisternas, L.A.; Jeldres, R.I.; González-Benito, G.; Rivas, M. Partial desalination of seawater for mining processes through a fluidized bed bioreactor filled with immobilized cells of Bacillus subtilis LN8B. *Desalination* **2020**, *482*, 114388. [CrossRef]
44. Nesset, J.E.; Hernandez-Aguilar, J.R.; Acuna, C.; Gomez, C.O.; Finch, J.A. Some gas dispersion characteristics of mechanical flotation machines. *Miner. Eng.* **2006**, *19*, 807–815. [CrossRef]
45. Lazić, Ž.R. *Design of Experiments in Chemical Engineering—A Practical Guide*; Wiley-Vch Verlag GmbH & Co. KGaA: Weinheim, Germany, 2004; ISBN 3527311424.
46. Arroug, L.; Elaatmani, M.; Zegzouti, A. A preliminary study to investigate the beneficiation of low-grade phosphate sludge using reverse flotation: Modeling and optimization through Box-Behnken design and response surface methodology. *Chem. Eng. Res. Des.* **2024**, *204*, 228–237. [CrossRef]
47. Jansen, M.J.W. Analysis of variance designs for model output. *Comput. Phys. Commun.* **1999**, *117*, 35–43. [CrossRef]
48. Lucay, F.A.; Lopez-Arenas, T.; Sales-Cruz, M.; Gálvez, E.D.; Cisternas, L.A. Performance profiles for benchmarking of global sensitivity analysis algorithms. *Rev. Mex. Ing. Quim.* **2020**, *19*, 423–444. [CrossRef]
49. Lucay, F.A.; Gálvez, E.D.; Salez-Cruz, M.; Cisternas, L.A. Improving milling operation using uncertainty and global sensitivity analyses. *Miner. Eng.* **2019**, *131*, 249–261. [CrossRef]
50. Sepúlveda, F.D.; Cisternas, L.A.; Gálvez, E.D. The use of global sensitivity analysis for improving processes: Applications to mineral processing. *Comput. Chem. Eng.* **2014**, *66*, 221–232. [CrossRef]
51. Corona-Arroyo, M.A.; López-Valdivieso, A.; Laskowski, J.S.; Encinas-Oropesa, A. Effect of frothers and dodecylamine on bubble size and gas holdup in a downflow column. *Miner. Eng.* **2015**, *81*, 109–115. [CrossRef]
52. Jávor, Z.; Schreithofer, N.; Heiskanen, K. Validity of critical coalescence concentration in dynamic conditions. *Int. J. Miner. Process.* **2014**, *127*, 16–22. [CrossRef]
53. Chen, X.; Peng, Y.; Bradshaw, D. Effect of regrinding conditions on pyrite flotation in the presence of copper ions. *Int. J. Miner. Process.* **2013**, *125*, 129–136. [CrossRef]
54. Mankosa, M.J.; Kohmuench, J.N.; Eisenmann, M.D.; Luttrell, G.H. *Testing of the Hydrofloat Separator for Coal Cleaning Applications*; Virginia Polytechnic Institute and State University: Blacksburg, VA, USA, 1995; pp. 1–6.

minerals

MDPI

Article

Open-Circuit Technology of Zinc Oxide Ore Flotation with Ternary Collector and Its Adsorption Characteristics on Smithsonite Surface

Zhiwei Li [1,2], Qicheng Feng [1,2], Qian Zhang [1,2,*] and Shuming Wen [1,2,*]

1 State Key Laboratory of Complex Nonferrous Metal Resources Clean Utilization, Faculty of Land Resource Engineering, Kunming University of Science and Technology, Kunming 650093, China; 18487207322@163.com (Z.L.); fqckmust@163.com (Q.F.)
2 Yunnan Key Laboratory of Green Separation and Enrichment of Strategic Mineral Resources, Kunming University of Science and Technology, Kunming 650093, China
* Correspondence: zqian9865@163.com (Q.Z.); shmwen@126.com (S.W.)

Abstract: The sulfidization-amine flotation method is commonly used for the beneficiation of zinc oxide ores. Lanping zinc oxide ores contains 8.40% zinc, with the main mineral being smithsonite; additionally, they have a high mud content. Conventional sulfidization–ammonium flotation presents a low flotation index and unsatisfactory flotation froth. A new open-circuit technology is employed to treat Lanping zinc oxide ores, where Na_2S, KG-248, and dodecyl amine + sodium isoamyl xanthate + ammonium dibutyl dithiophosphate are used as the regulator, depressant, and ternary collector, respectively. Consequently, the flotation indices for the zinc grade and recovery are 28.71% and 86.24%, respectively, and the flotation froth becomes more stable. Subsequently, the flotation behavior and adsorption mechanism of smithsonite with a ternary collector are investigated. The flotation recovery of smithsonite increases to 94.40% after treatment with the ternary collector. Surface-analysis results indicate that the ternary collector can synergistically adsorb onto the sulfidized smithsonite surface to enhance its hydrophobicity, thus increasing the floatability of smithsonite. Meanwhile, the total consumption of the collector in the ternary-collector system is lower than that in the binary- or unitary-collector system.

Keywords: open-circuit technology; zinc oxide ore; smithsonite; flotation; ternary collector

1. Introduction

Zinc is a crucial component in the field of non-ferrous metals and is widely used in various industries, such as the electrical, mechanical, military, metallurgical, chemical, optical, and pharmaceutical industries [1,2]. The primary source for zinc extraction is zinc sulfide ore; however, the continuous increase in global demand for zinc resources has resulted in the gradual depletion of zinc sulfide ore resources. Therefore, the possibility of using zinc oxide ores must be investigated to satisfy the increasing demand [3–5].

Flotation is the most commonly used method for enriching zinc oxide ores and includes direct flotation (fatty acid, flocculation, and chelation) and sulfidization flotation (sulfidization amine and sulfidation xanthate) [6–16]. Among them, sulfidization-amine flotation treated with zinc oxide ore has been widely employed and investigated owing to its high flotation recovery [17,18]. Feng et al. [19] discovered that the quality of the sulfidation effect determines whether zinc oxide can yield excellent results and that sulfidation can yield zinc monosulfide and zinc polysulfide. Additionally, adding an appropriate amount of the sulfiding agent can yield more polysulfides, thus exerting activation and promotion effects on the flotation of zinc oxide ores. However, excessive sulfidation causes the generation of excessive sulfide ions in the pulp, thereby inhibiting the flotation effect. Mehdilo et al. [20] discovered that the dosage of sulfidizing agents and the pulp pH significantly affected the effectiveness of sulfidization-amine flotation methods. In the flotation process of smithsonite, when the optimal dosage of sodium sulfide was 500 g/t and Armac C and Armac T were used as collectors, 82.60% and 83.70% recoveries were achieved, respectively. However, a continued increase in the sulfide concentration decreased the smithsonite recovery. Additionally, sulfidization-amine methods pose several severe issues, such as the generation of significant amounts of viscous froth when amine collectors are added. These viscous froths are not easily defoamed and are sensitive to the slurry, thus resulting in a significant decline in the flotation index and the emergence of empty and large bubbles [21–24].

To address these challenges, researchers have investigated combinations of anionic and cationic collectors for the treatment of zinc oxide ores [25–28]. Anionic or cationic collectors with opposite charges promote adsorption through electrostatic interactions. Hosseini et al. [29] discovered that the recovery of smithsonite was low regardless of the collector concentration when using a single collector, whereas the recovery of smithsonite increased significantly by employing potassium isoamyl xanthate (KAX) and dodecyl amine (DDA) combined collectors. This is attributable to the use of anion and cation collectors, which decreased the electrostatic head–head repulsion between the surface and ammonium ions and increased the lateral tail–tail hydrophobic bonds. Majid et al. [30] employed Armac C and KAX as anionic and cationic collectors, respectively, to obtain zinc oxide minerals synergistically. Sulfidized regions on the zinc oxide surface with a negative charge were termed "strong spots". Positively charged Armac C ions were firmly adsorbed onto these "strong spots", whereas negatively charged KAX ions were easily adsorbed with Armac C, thereby increasing the surface coverage of the collectors on the mineral surface and enhancing the floatability. Wang et al. [31] mixed sodium oleate and butyl hydroxy anisole in a molar ratio of 2:1 to prepare a new combined collector named "BHOA". It can adsorb onto the surface of smithsonite, thus reducing the surface potential to below −50 mV. Consequently, the sulfidization of smithsonite is improved, thus enhancing its flotation performance. Furthermore, Wang et al. [32] discovered that by combining KAX and DDA for smithsonite flotation, DDA ions occupied the space between KAX anions, thereby preventing adjacent KAX from experiencing electrostatic repulsion. This increased the attraction between the tail groups, thus allowing the further adsorption of DDA cations on the surface of smithsonite; consequently, a remarkable concentrate recovery of 96.80% was achieved. Combining anionic and cationic collectors to treat zinc oxide minerals enhances the hydrophobicity of the mineral surface, owing to the synergistic effects between the agents. In the alkaline pH range, where the mineral surface is negatively charged, cations are preferentially adsorbed onto the mineral surface, which are subsequently co-adsorbed through electrostatic neutralization, thus resulting in complexes that enhance the floatability of the mineral [33,34]. However, in industrial production for low-grade zinc oxide ores characterized by a fine particle size, a high mud content, and abundant gangue minerals, the floatability difference between gangue and purpose minerals is minimal. Furthermore, gangue reacts easily with the amine collector, thus deteriorating the concentrate grade and recovery [35–38]. Therefore, the appropriate combination of anionic and cationic collectors that exhibits synergistic effects must be identified.

In this work, owing to its low grade, fine particle size, and more kinds of gangue minerals, the Lanping zinc oxide ore renders sulfidization-amine flotation ineffective. After optimizing the reagent system and flotation process flow, a ternary-collector system (DDA + NaIX + ADD) was used to treat zinc oxide ore. Subsequently, a new open-circuit process and depressant (KG-248) were introduced to minimize the effects of gangue, slimes, and residual froth in the pulp circulation on the flotation concentrate. Additionally, microflotation, zeta-potential analysis, adsorption tests, X-ray photoelectron spectroscopy (XPS), and contact-angle measurements were performed to elucidate the surface mechanisms of the cooperative adsorption of the ternary collector on the smithsonite surface. This study offers theoretical support for the development and utilization of zinc oxide ores with a high mud content that cannot be beneficiated easily.

2. Experiment

2.1. Materials and Reagents

Zinc oxide ore was obtained from Jinding Zinc Industry Co., Ltd., Lanping County, Yunnan Province, China. Samples smaller than 2 mm were obtained by crushing and sieving. The samples were ground to −0.074 μm and then prepared for analysis after being dried. Results of chemical multi-element analysis revealed that the zinc and lead grades in the ore were 8.40% and 1.26%, respectively. The contents of calcite and quartz in the raw ore are 35.33% and 26.26%, respectively, which are the main components of gangue minerals. Meanwhile, results of phase analysis indicated that the zinc appeared primarily in the form of smithsonite, which constituted 77.99% of the total element. The occurrence form and distribution rate of the remaining zinc are, respectively, 11.53% for sphalerite, 7.74% for hemimorphite, 2.63% for limonite, 0.10% for siderite, and 0.01% for rhodochrosite. According to the distribution rate of zinc in the ore, smithsonite should be considered in the recovery.

Smithsonite samples were obtained from Lanping County, Yunnan Province, China. A high-purity smithsonite sample was obtained via meticulous selection, after which it was cleaned ultrasonically and dried naturally. Subsequently, it was ground using an agate grinder, following which smithsonite samples of different sizes were obtained via screening using a Taylor standard sieve. Finally, microflotation experiments and analytical tests were conducted. The X-ray diffraction (XRD) results of the smithsonite sample, as shown in Figure 1, indicated no clear impurity peaks. Meanwhile, the elemental-analysis results of the smithsonite sample indicated that the zinc content was 50.14%. These results confirmed the high purity of the smithsonite sample, which satisfies the requirements of a single-mineral flotation experiment.

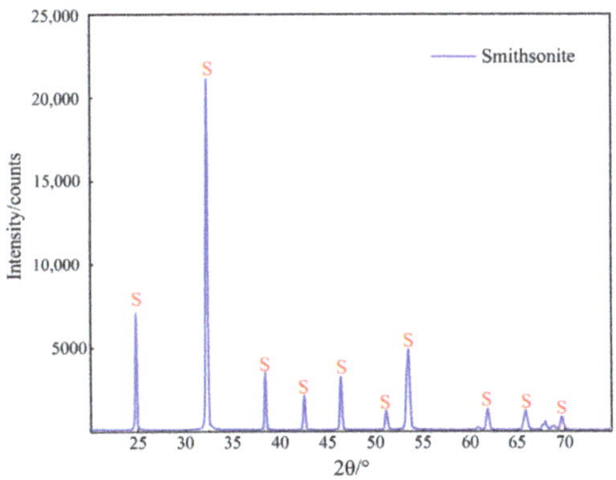

Figure 1. XRD pattern of smithsonite sample.

Commercially pure Na_2CO_3, Na_2S, KG-248, NaIX, ADD, and $2^{\#}$ oil and analytically pure DDA were used in the flotation tests of the zinc oxide ores. Na_2CO_3 was used to regulate the lead-flotation process. Na_2S was used to sulfidize lead and zinc oxide, whereas KG-248 was used as a depressant in the zinc flotation process. NaIX, DDA, and ADD were used as collectors, and the combination of NaIX and ADD collectors was named GS-2. For the flotation of smithsonite, $Na_2S{\cdot}9H_2O$ (analytically pure), ADD (analytically pure), NaIX (commercially pure), ADD (commercially pure) were used as the sulfidization reagent (source of S^{2-} ions), collector, and frother, respectively. Additionally, NaOH (analytically pure) and HCl (analytically pure) were utilized as pH regulators, and NaCl (analytically pure) was utilized as the electrolyte for zeta-potential measurements. Tap water and deionized water were used in the zinc oxide ore and smithsonite tests, respectively. Notably, KG-248 was a macromolecular organic agent with a molecular weight of about 20,000, which was a self-made agent with good water solubility, no highly toxic and harmful components, and had a good effect on inhibiting carbonate and silicate gangue minerals.

2.2. Flotation Tests

Flotation was conducted on the zinc oxide ores using an XFD flotation machine. Specifically, 400 g of zinc oxide ore with a particle size of less than 2 mm and 400 mL of tap water were added to a rod mill to obtain ground products with the desired fineness. The grinding fineness was obtained by collecting the screen undersize using the Taylor standard screen. Subsequently, the ground products were introduced into the cell of the 1.5 L XFD flotation machine, followed by the sequential addition of Na_2CO_3, Na_2S, NaIX, and $2^{\#}$ oil, which were stirred for 3, 5, 3, and 2 min, respectively. The lead rough concentrate was obtained after lead flotation. Subsequently, Na_2S, KG248, DDA (dodecyl amine), NaIX (sodium isoamyl xanthate), and ADD (ammonium dibutyl dithiophosphate) were added to the lead tailing pulp and reacted for 5, 3, 3, 3, and 3 min, respectively, and the zinc rough concentrate was obtained after zinc flotation. The obtained products were dried, weighed, and Atomic Absorption Spectroscopy was used for analysis; subsequently, the recovery was calculated. After determining the reagent dosage, open- and closed-circuit tests were performed based on a specified flow chart.

In the smithsonite microflotation test, 2 g of smithsonite with a particle size of −75 to +38 μm and deionized water (35 mL) were introduced into a 40 mL flotation cell. The pulp pH was adjusted to 11 using NaOH. Subsequently, Na_2S, DDA, NaIX, and ADD solutions were added and stirred for 5, 3, 3, and 3 min, respectively. Subsequently, the froth (concentrate) and cell products (tailing) were obtained, and the flotation recovery of smithsonite was calculated after it was dried and weighed.

2.3. Zeta Potential

In this study, zeta potentials were measured using a Zetasizer-3000HS instrument (Malvern, Worcestershire, UK). The prepared smithsonite sample (20 mg, −5 μm) was placed inside a beaker containing NaCl solution (1 mM), the pulp pH was adjusted, and reagents were added and stirred for a certain duration depending on the flotation-test requirements. After the reaction was completed, the mixture was allowed to settle naturally for 10 min, and the supernatant was extracted using a disposable plastic straw to test the zeta potential. Each sample was measured three times, and the average value was recorded as the final value.

2.4. Adsorption Tests

Ultraviolet–visible spectroscopy was performed to evaluate the adsorption capacity of the collector on the smithsonite surface. Initially, 2 g of smithsonite (−38 μm) was introduced into a beaker containing 35 mL of deionized water. The necessary reagents were added and stirred for a certain duration using a magnetic stirrer, depending on the flotation-experiment requirements. After the reaction was completed, the mixture was allowed to precipitate. Subsequently, the supernatant was extracted using a disposable plastic straw and then transferred to a centrifuge tube. After centrifugation for 15 min, the supernatant was transferred meticulously to a quartz cuvette and analyzed using a UV-2700 spectrophotometer (Shimadzu, Kyoto, Japan). The collector concentration remaining in the solution was obtained, and the amount of collector adsorbed onto the smithsonite surface was calculated. For NaIX and ADD, UV spectroscopy was used to directly measure the residual concentration in the supernatant. Due to the lack of characteristic absorption peaks in the ultraviolet spectrum of DDA, eosin-Y was used as an indicator for the determination of DDA concentration, and the specific steps were consistent with those in the previous literature [39].

2.5. XPS Analysis

XPS analysis was performed to qualitatively and quantitatively assess the elemental composition and chemical state of the smithsonite sample. Specifically, 2 g of smithsonite (−38 μm) was mixed with 35 mL deionized water in a beaker, followed by the addition of the required reagents after the pulp pH was adjusted. The smithsonite and reagent reacted sufficiently after stirring was performed for a certain duration using a magnetic stirrer. Subsequently, the mixture product was filtrated. The filtered residue was rinsed with deionized water and then allowed to dry naturally. Subsequently, it was subjected to XPS analysis using a pHI5000 spectrometer (ULVAC-PHI Inc., Kanagawa, Japan).

2.6. Contact-Angle Test

Contact-angle measurements were performed using a JY-82 contact-angle analyzer. The rectangular smithsonite sample was polished using 400-grit and 5000-grit aluminum-oxide sandpaper and cleaned using ethanol. Subsequently, it was placed in a beaker containing 35 mL of deionized water, the pulp pH was adjusted to 11, and the required reagents were added in the specified sequence. The sample was removed after the reagents reacted with the mineral surface. Next, it was rapidly dried and transferred to a platform. Subsequently, its contact angle was measured using the liquid-droplet method.

3. Results and Discussion

3.1. Flotation of Zinc Oxide Ores

To weaken the effect of lead on the flotation recovery of zinc, we prioritized the recovery of lead from zinc oxide ores. The optimal reagent conditions were 1000 g/t Na_2CO_3, 2000 g/t Na_2S, 250 g/t NaIX, and 40 g/t $2^{\#}$ oil, which resulted in 7.16% and 80.40% for the lead concentrate grade and recovery, respectively. Subsequently, the flotation of zinc, including the conditions and open-circuit tests, was conducted to increase the flotation index of the zinc concentrate.

3.1.1. Flotation Test of Zinc

Grinding fineness as well as reagent type and dosage are important parameters affecting the mineral flotation index. Based on the properties of zinc oxide ores, many exploratory experiments have been conducted to determine the reagent types for conducting zinc flotation. Ultimately, Na_2S, KG-248, and the combination of DDA + NaIX + ADD were identified as the sulfidization reagent, depressant, and ternary collector, respectively, to recover zinc. Subsequently, the effects of the grinding fineness and reagent dosage on zinc flotation were investigated, and the results are presented in Figures 2 and 3.

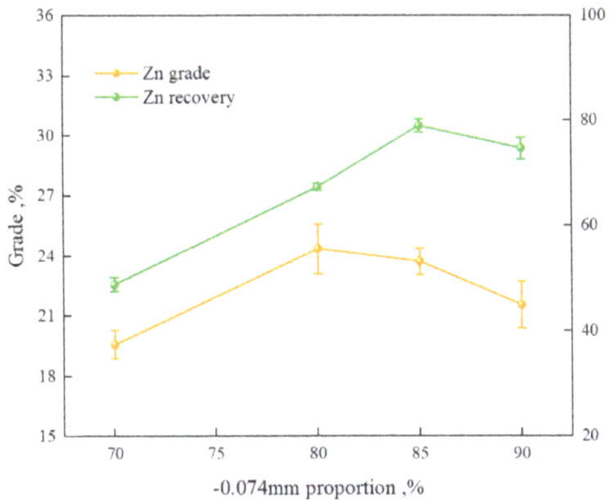

Figure 2. Effect of grinding fineness on zinc flotation.

As shown in Figure 2, the grinding fineness of the ores affected the grade and recovery of the zinc rough concentrate. The grade and recovery of zinc first increased and then decreased. The best indices were achieved when 85% of the particles exhibited a grinding fineness of −0.074 mm. Meanwhile, the zinc grade and recovery were 23.74%, and 79.22%, respectively, when less than 85% of the particles measured −0.074 mm, which is attributable to the insufficient dissociation of the purpose mineral. By contrast, the ore was overground when the proportion exceeded 85%, which exacerbated the mud problem and thus adversely affected the grade and recovery index.

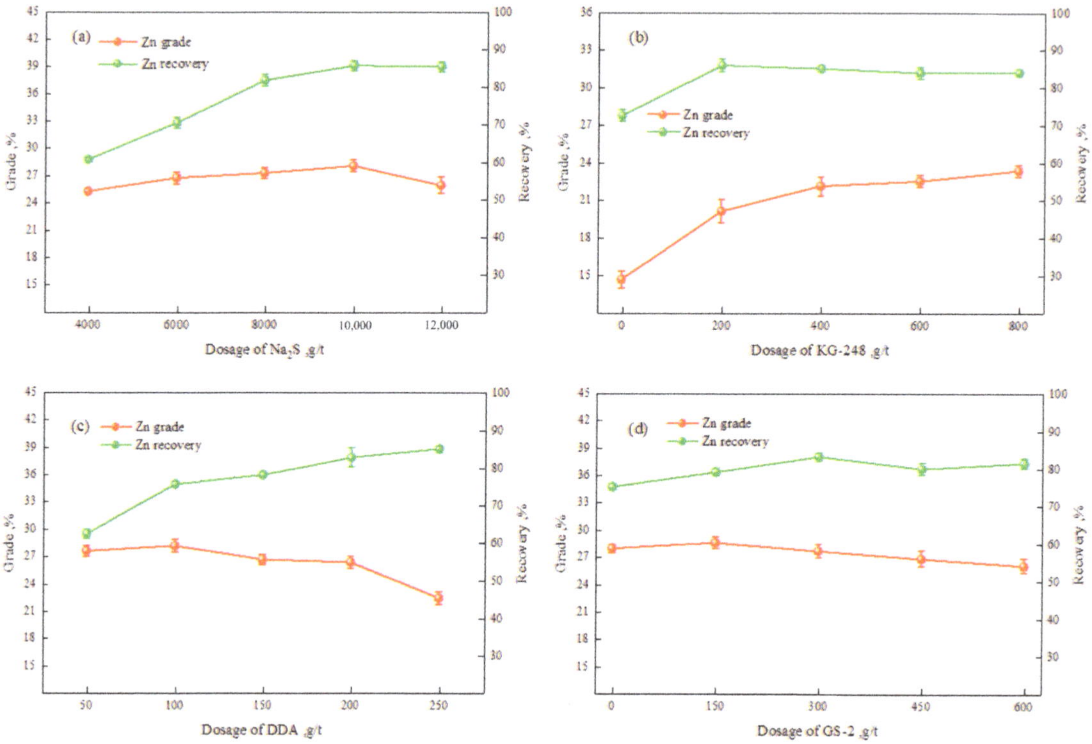

Figure 3. Effect of reagent dosage on zinc flotation: (**a**) Na_2S; (**b**) KG-248; (**c**) DDA; (**d**) GS-2.

Na_2S would produce HS^- and S^{2-} ions, which could be firmly adsorbed on the mineral surface and form a film with ZnS as the main component [19]. This film made the mineral easier to combine with the flotation agent, which not only enhanced the surface hydrophobicity of the mineral but also facilitated the smooth progress of the subsequent flotation process. Because the C=S bond on the functional group of the thiol collector was very close to the polarity of the metal sulfide. This similarity allowed them to form a tight combination to form a MeS (Me: metal element) cell [14]. These crystal cells were like a bridge connecting the collector and the surface of the oxidized mineral, ensuring that the adsorption of the collector on the mineral surface could meet the needs of froth mineralization. As shown in Figure 3a, the grade and recovery of zinc initially increased and then decreased as the Na_2S dosage increased from 4000 to 12,000 g/t. The optimal flotation index was obtained when the Na_2S dosage was 10,000 g/t, thus indicating that a minute amount of Na_2S was insufficient to completely sulfidize the zinc oxide surface, thus resulting in an undesirable flotation index. However, excessive Na_2S suppressed the flotation of zinc oxide. This is attributed to the residual Na_2S in the pulp solution, which resulted in competitive adsorption with the collector. Compared with the case without KG-248, the grade and recovery of zinc increased significantly when KG-248 was added (Figure 3b), thus indicating that KG-248 significantly affects dispersed mud and gangue minerals. However, the method for improving the zinc flotation index when the dosage of KG-248 exceeds 400 g/t remains ambiguous; therefore, the dosage of KG-248 was set as 400 g/t.

Figure 3c shows that although the zinc recovery increased gradually, the grade decreased as the DDA dosage increased from 50 to 250 g/t. This is because as the DDA dosage increased, DDA was adsorbed onto the surface of mud and gangue minerals, which resulted in the formation of many froth layers. Consequently, the zinc grade decreased and the flotation environment deteriorated. Some studies demonstrated that combining anions and cations as collectors effectively enhanced the stability of froth [40]. Therefore, an anion collector (GS-2) was added to effectively obtain zinc in the cation (DDA) system, and the results are shown in Figure 3d. When the DDA dosage was 200 g/t, the grade and recovery of zinc increased with the addition of 150 g/t GS-2, the flotation environment improved, and the flotation froth layer became stable; however, when the dosage of GS-2 was increased to 300 g/t, zinc recovery increased significantly but the grade deteriorated slightly. Thus, the addition of 300 g/t anionic collector to the DDA system enhanced the flotation index and improved flotation, thus facilitating the beneficiation of zinc.

3.1.2. Flotation Open-Circuit Test

After determining the optimal reagent and dosage via condition tests, the process optimization of the zinc oxide ore was carried out in order to maximize the recovery of the purpose mineral. Ultimately, a new open-circuit flotation process was performed to treat the zinc oxide ores, and verification tests were conducted based on the flow chart presented in Figure 4. Adopted the flotation process of the first lead beneficiation, followed by that of zinc. The zinc concentration was measured after one rougher step, one scavenger step, and two cleaner steps. To avoid the deterioration of the flotation index and the phenomenon caused by the sequential return of middling ore as well as by the middling ore arising from lead cleaner and zinc cleaner I, merged and treated the concentrates obtained from zinc scavenger in zinc cleaner II. The feasibility and stability of the open-circuit process were verified five times, and a favorable and stable flotation index was obtained, as listed in Table 1.

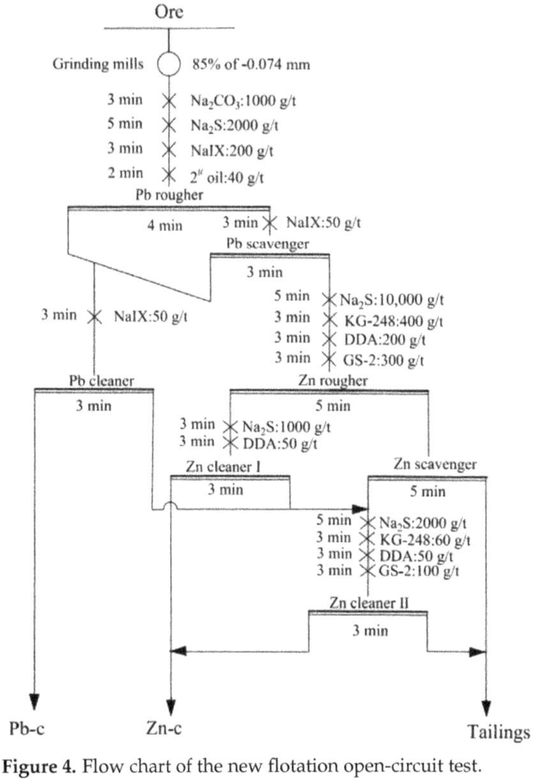

Figure 4. Flow chart of the new flotation open-circuit test.

Table 1. Results of the new flotation open-circuit test.

Product	Yield, %	Zn Grade, %	Zn Recovery, %
Pb-c	4.89	4.96	2.97
Zn-c	24.52	28.71	86.24
Tailings	70.59	1.25	10.79
Sum	100.00	8.16	100.00

The data in Table 1 show that qualified zinc concentrate products were obtained via the new open-circuit test. The zinc concentrate grade and recovery were 28.71% and 86.24%, respectively. Furthermore, in the absence of closed-circuit processes, using the new open-circuit process obtained more advantages. The combination of anionic and cationic collectors could enhance the collection effect so that fewer collectors could be used to obtain a better flotation effect. A reasonable dosage of DDA and an open-circuit process that could discharge ore in time would effectively solve the problem of froth accumulation in flotation cells.

The cost of the reagents used in the process in Figure 4 was estimated to ensure its economic viability. The data of each reagent are taken from China, as listed in Table 2.

Table 2. Economic estimates for reagents.

Name	Price, ¥/t	Dosage, g/t	Price, ¥
Na_2CO_3	800	1000	0.8
Na_2S	2200	15,000	33
NaIX	10,000	300	3
KG-248	15,500	460	7.13
DDA	17,800	300	5.34
GS-2	12,500	400	5
$2^{\#}$ oil	6000	40	0.24
Sum	-	-	54.51

As can be seen from Table 2, the cost per tonne of ore treated with the reagents regime is approximately 54.51¥, and the zinc concentrate with a grade of more than 28.00% and a recovery of more than 85% can be obtained after treatment by this flotation process. Notably, there was no return stage in the open-circuit process, which meant that less equipment and energy consumption would be used, which contributed to energy saving and environmental protection. Thus, this approach is promising when used in the industrial application of zinc oxide ores.

3.2. *Flotation Behavior and Surface-Adsorption Analysis of Smithsonite*

The flotation results of the zinc oxide ore indicated that the open-circuit process with ternary-collector treatment (DDA + NaIX + ADD) yielded a stable flotation froth and an exceptional flotation index. The phase-analysis result of the raw ore shows that the zinc oxide ore appeared primarily in the form of smithsonite. Therefore, we investigated the effect of the ternary collector (DDA + NaIX + ADD) on the smithsonite flotation behavior as well as the adsorption mechanism on the smithsonite surface.

3.2.1. Effect of Ternary Collector on Smithsonite Flotation Behavior

The effects of pH and Na_2S concentration on the floatability of smithsonite are shown in Figure 5. Based on Figure 5a, as the pH increased from 5 to 11, the floatability of smithsonite increased, and its recovery increased from 23.20% to 64.90%. However, its flotation recovery decreased slightly as the pH increased to 12, thus indicating that the optimum flotation pH for smithsonite was 11. The effect of Na_2S concentration on the smithsonite floatability in different collector systems at pH 11 is depicted in Figure 5b; the recovery of smithsonite improved as the Na_2S concentration increased from 1×10^{-4} to 8×10^{-4} mol/L. The optimum flotation effectiveness for smithsonite was reached at the dosage of Na_2S of 8×10^{-4} mol/L. Beyond the concentration of 8×10^{-4} mol/L, the amount of Na_2S continued to increase, but the recovery did not improve significantly; there was little benefit under all three systems. Notably, the excess Na_2S did not significantly inhibit the co-adsorption of the collectors (DDA, NaIX, and ADD) on the surface of smithsonite under the ternary-collector system, which signifies a favorable synergistic effect.

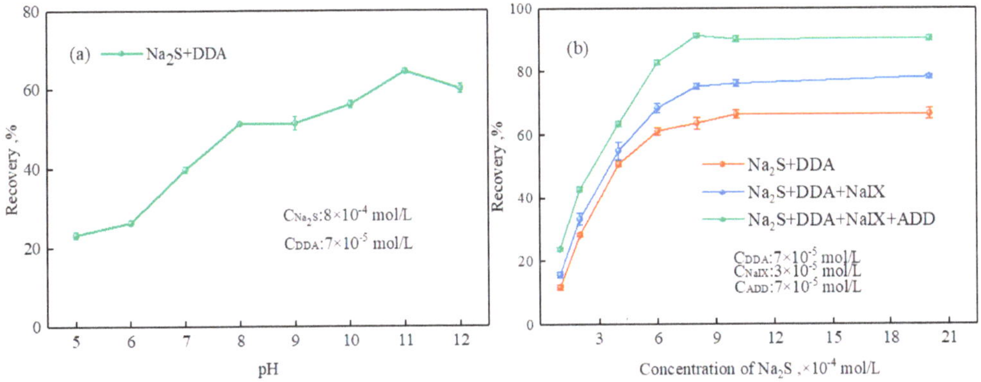

Figure 5. Effects of (**a**) pH and (**b**) Na_2S concentration on smithsonite floatability.

Figure 6 shows the effect of the collector concentration on the smithsonite floatability in different systems. The floatability of the sulfidized smithsonite after the ternary-collector treatment was superior to that after the binary- and unitary-collector treatments. At low concentrations, DDA ($<7 \times 10^{-5}$ mol/L) did not significantly improve the recovery. However, at a DDA concentration of 7×10^{-5} mol/L, the recovery of smithsonite increased significantly. A further increase in DDA concentration did not significantly enhance the flotation effect (Figure 6a). In the binary-collector system, the addition of NaIX significantly improved the recovery of smithsonite, with better performances demonstrated at lower dosages, and the recovery increased to 74.80% (Figure 6b). In the DDA + ADD collector system (Figure 6c), ADD exhibited excellent acquisition ability, thus increasing the smithsonite recovery to 76.20% at a concentration of 7×10^{-5} mol/L. In the ternary-collector system, the maximum smithsonite recovery achieved was 94.40% (Figure 6c), which was much higher than those achieved by the binary and single collectors. Notably, the co-adsorption of the ternary collector on the smithsonite surface did not result in mutual inhibition. By contrast, collectors with opposite charges enhanced the co-adsorption effect, allowing more collectors to be adsorbed onto the smithsonite surface, thus improving the surface hydrophobicity of smithsonite. Additionally, in the presence of both anionic collectors, the half-life of the froth generated by DDA reduced significantly.

3.2.2. Zeta-Potential Measurements

The zeta potential is an important parameter for assessing the electrical properties of mineral particles at the solid–liquid interface [41,42]. The zeta potential of the sulfidized smithsonite surface under various collector systems was measured, and the results are shown in Figure 7. As shown, the zeta potential shifted in the negative direction as the pH increased from 5 to 12, which is consistent with the results reported in the literature [43]. Compared with the case of untreated sulfidized smithsonite, the zeta potential shifted in the positive direction after DDA, DDA + NaIX, and DDA + NaIX + ADD treatments, thus indicating the adsorption of positively charged ammonia species on the sulfidized smithsonite surface. However, the zeta potential shifted in the negative direction in the DDA + NaIX system as compared with that in the DDA system, which is attributable to the adsorption of negatively charged species on the sulfidized smithsonite surface. Furthermore, the addition of ADD to the pulp solution in the DDA + NaIX system resulted in a more negative zeta potential, which is attributable to the adsorption of dibutyl dithiophosphate species on the sulfidized smithsonite surface. These phenomena indicate that the ternary collector was adsorbed onto the sulfidized smithsonite surface, which enhanced the hydrophobicity of the mineral surface and benefited the flotation of smithsonite.

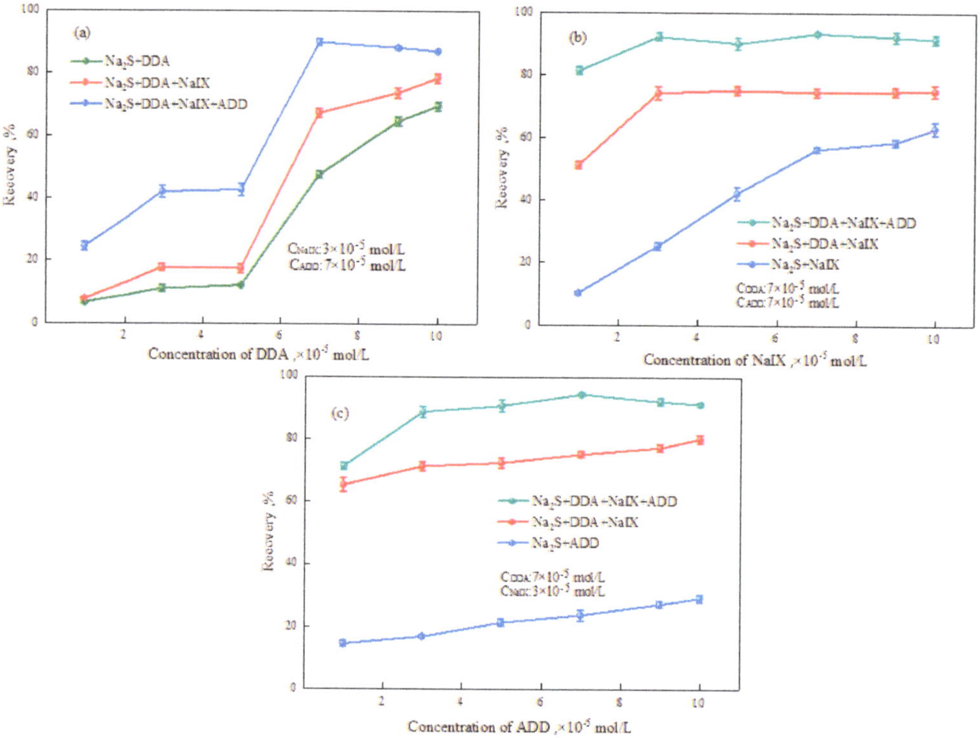

Figure 6. Effect of collector concentration on smithsonite floatability: (**a**) DDA; (**b**) NaIX; (**c**) ADD.

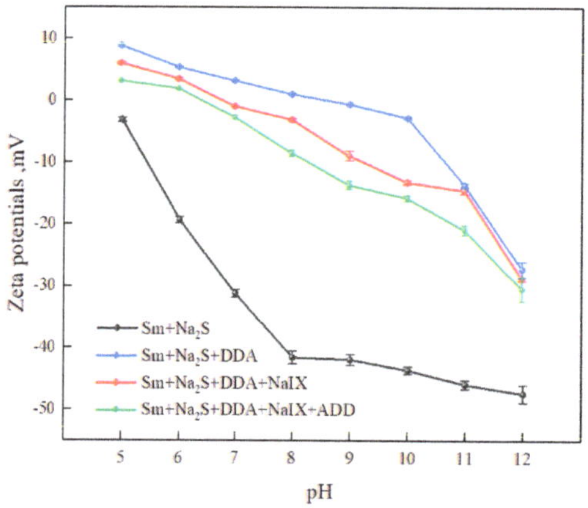

Figure 7. Zeta potentials of smithsonite (Sm) under different collector systems.

3.2.3. Adsorption Experiments

The adsorption behavior of the ternary-collector system on the sulfidized smithsonite surface was investigated via adsorption measurements. The initial collector concentration was denoted as C_0, the residual amount of the collector in the pulp C_1 was determined, and the adsorption capacity of the collector was calculated using the formula $\tau = v(C_0 - C_1)/m$ (Figure 8). Figure 8a illustrates the changes in the adsorption capacity of DDA in the different systems. Within the DDA concentration range of 1×10^{-5} to 5×10^{-5} mol/L, the adsorption capacities of all three systems varied only slightly, thus indicating low adsorption efficiency at low DDA concentrations. However, the adsorption of DDA increased significantly as the DDA concentration increased to 7×10^{-5} mol/L, particularly under the ternary-collector system. Meanwhile, the adsorption capacity increased from 5.31×10^{-6} to 9×10^{-7} mol/g, thus suggesting that the adsorption of NaIX and ADD enhanced the adsorption of DDA on the sulfidized smithsonite surface. Similarly, as shown in Figure 8b,c, the adsorption capacity of NaIX and ADD increased with the amounts of NaIX and ADD. In particular, the ternary system exhibited a greater increase than the binary and unitary systems, thus indicating that the ternary collector not only promoted adsorption but also synergistic adsorption, thus resulting in the absorption of more collectors onto the sulfidized smithsonite surface, which consequently enhanced its floatability.

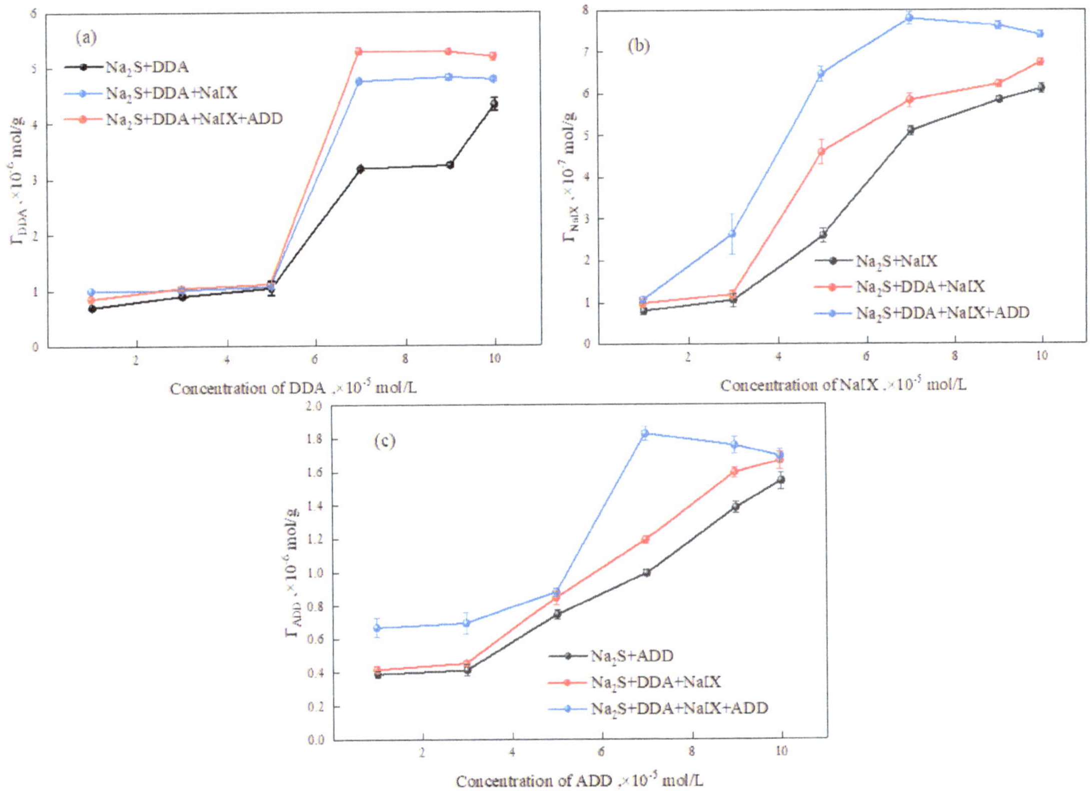

Figure 8. Adsorption capacity of collectors in different systems vs. concentrations of (**a**) DDA, (**b**) NaIX, and (**c**) ADD.

3.2.4. XPS Analysis

To further examine the synergistic adsorption mechanism of the ternary collector on the sulfidized smithsonite surface, the elemental composition and chemical state of the mineral surface were analyzed using XPS [44,45]. Figures 9 and 10 illustrate the elemental composition and high-resolution N 1s XPS spectra of the smithsonite surface after treatment with different reagents.

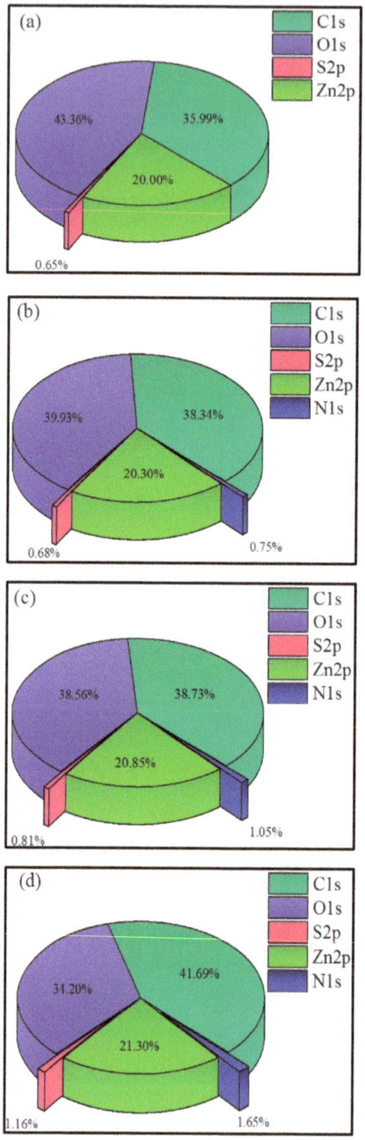

Figure 9. Proportion of elements on smithsonite surface treated with (**a**) Na_2S, (**b**) Na_2S + DDA, (**c**) Na_2S + DDA + NaIX, and (**d**) Na_2S + DDA + NaIX + ADD.

Only C, O, S, and Zn were detected on the sulfidized smithsonite surface (Figure 9a) at proportions of 35.99%, 43.36%, 0.65%, and 20.00%, respectively. However, with the addition of DDA (Figure 9b), a new N element was discovered on the sulfidized smithsonite surface at a proportion of 0.75%. Identically, the data in Figure 10a show that no N 1s peak was detected on the sulfidized smithsonite surface. Additionally, with the addition of DDA (Figure 10b), a new N 1s XPS spectrum comprising two split peaks with binding energies of 399.94 and 405.39 eV were detected on the mineral surface, which corresponded to -NH_2 and R-NH_3^+ groups in the DDA species, respectively [46,47]. Meanwhile, the distributions of -NH_2 and R-NH_3^+ were 67.43% and 32.57%, respectively, thus indicating that DDA was adsorbed onto the mineral surface. After DDA + NaIX treatment, the binding energies of -NH_2 and R-NH_3^+ groups were 399.90 and 405.40 eV, respectively; the distribution of -NH_2 increased from 67.43% to 73.68%, the proportion of R-NH_3^+ decreased by 6.25% (Figure 10c), and the proportion of N increased to 1.05% (Figure 9c). This shows that a higher amount of DDA was transferred to the mineral surface from the pulp solution and that the presence of NaIX enhanced the adsorption of DDA on the sulfidized smithsonite surface. In the DDA + NaIX + ADD ternary-collector system (Figure 10d), a new peak appeared at a binding energy of 402.02 eV, which originated from the C-N species in the ADD molecules, thus indicating the adsorption of ADD on the sulfidized smithsonite surface. Additionally, the increase in the peak area and intensity indicates that the adsorption of DDA increased, thus resulting in an increase N to 1.65% (Figure 9d). These results indicate a more intense adsorption reaction on the sulfidized smithsonite surface under the synergistic action of the DDA + NaIX + ADD ternary-collector system. The XPS analysis results are consistent with the test results of the zeta potential and adsorption capacity.

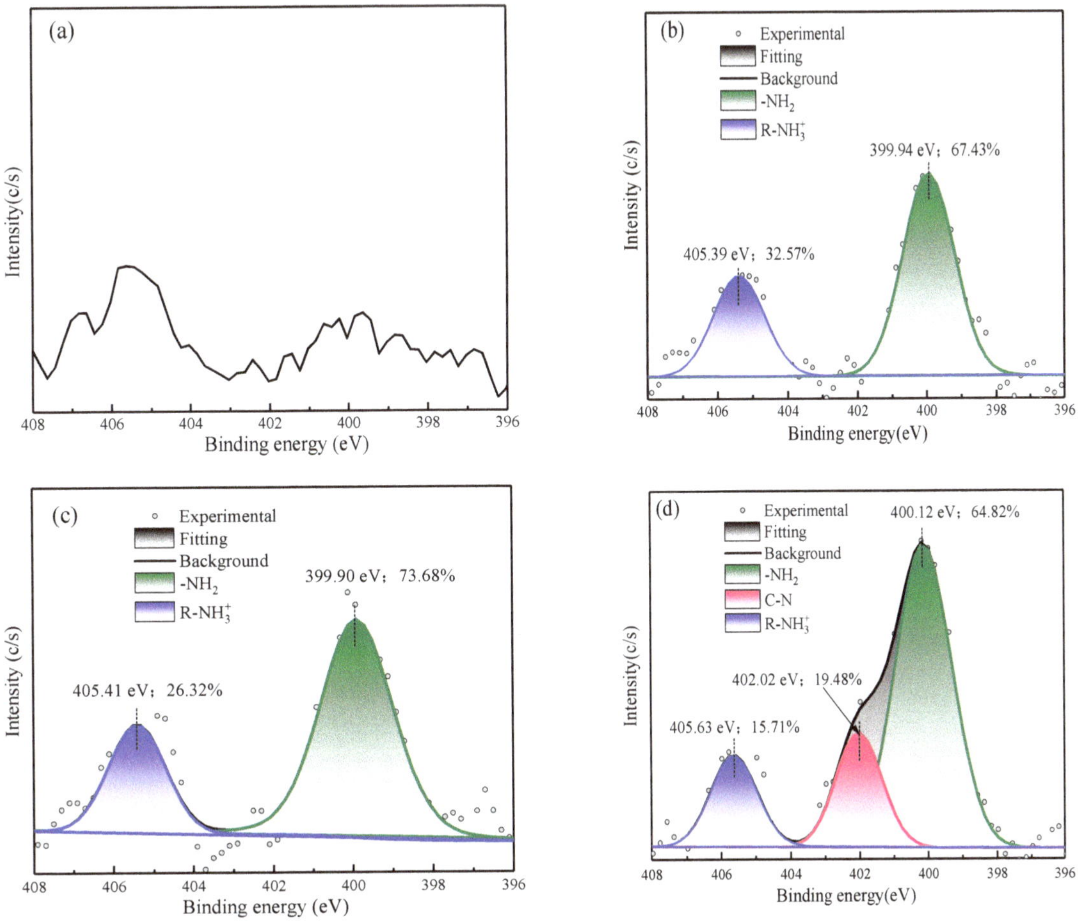

Figure 10. N 1s XPS spectra of smithsonite surface treated with (**a**) Na_2S; (**b**) Na_2S + DDA; (**c**) Na_2S + DDA + NaIX; and (**d**) Na_2S + DDA + NaIX + ADD.

3.3. Contact-Angle Test

The contact angles of the smithsonite surface under various conditions are shown in Figure 11. The contact angle of the untreated smithsonite surface was 33.12° (Figure 11a), which is consistent with the result in the literature [48], thus indicating that the smithsonite surface is highly hydrophilic. The contact angle of the sulfidized smithsonite surface treated with DDA increased to 60.17° (Figure 11b), thus indicating that DDA adsorbed onto the sulfidized smithsonite surface and increased the hydrophobicity of the mineral surface. In the binary-collector system (DDA + NaIX), the contact angle further increased to 62.92° (Figure 11c). Based on the result above as well as the adsorption capacity and XPS analysis, the addition of NaIX improved the adsorption of DDA on the mineral surface. After treatment with the ternary-collector system (DDA + NaIX + ADD), the contact angle of the sulfidized smithsonite surface increased significantly to 79.05° (Figure 11d), which may have promoted the synergistic adsorption of the ternary collector, thus enhancing the hydrophobicity of the mineral surface and improving the floatability of the sulfidized smithsonite.

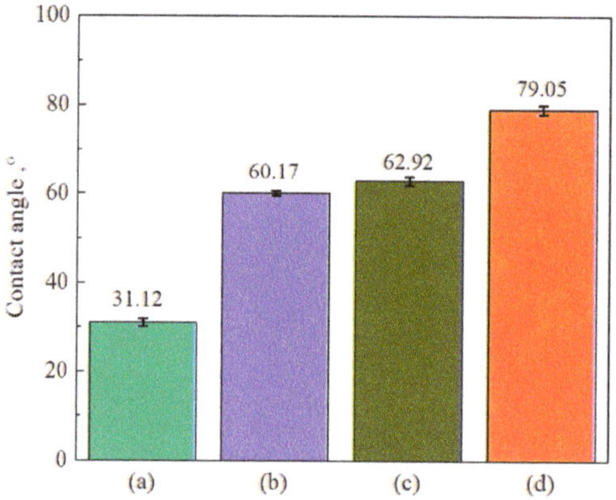

Figure 11. Contact angles of smithsonite surface treated with (**a**) deionized water, (**b**) Na_2S + DDA, (**c**) Na_2S + DDA + NaIX, and (**d**) Na_2S + DDA + NaIX + ADD.

4. Conclusions

In this study, a new open-circuit process using a ternary-collector system (DDA + NaIX + ADD) was developed to treat low-grade refractory zinc oxide ore. The floatability of smithsonite and the synergistic adsorption mechanism of the ternary collector on the sulfidized smithsonite surface were investigated via flotation experiments and surface analysis. The following conclusions were obtained:

1. By adopting the new open-circuit process and using a ternary-collector system to treat low-grade zinc oxide ore, favorable flotation indices of 28.71% and 86.24% were obtained for the zinc grade and recovery, respectively. Additionally, the open-circuit process effectively solved the problem of froth buildup caused by amine collectors during flotation.
2. The results of microflotation experiments revealed that the floatability of smithsonite enhanced significantly by employing the ternary-collector treatment. Specifically, the recovery of smithsonite increased to 94.40%.
3. The zeta-potential measurements and adsorption-capacity test results indicated that the ternary collector can be co-adsorbed onto the sulfidized smithsonite surface and thus reduce the total consumption of reagents.
4. XPS results showed that the addition of NaIX promoted DDA adsorption on the mineral surface and that the adsorption of ADD increased the contact angle of the sulfidized smithsonite surface treated with the ternary collector, thus enhancing the hydrophobicity and floatability of the smithsonite.

Author Contributions: Conceptualization, S.W.; writing—original draft, Z.L.; writing—review and editing, Q.Z.; supervision, Q.F. All authors have read and agreed to the published version of the manuscript.

Funding: This research was funded by Yunnan Science and Technology Leading Talent Project grant number 202305AB350005.

Data Availability Statement: All relevant data are within the manuscript.

Conflicts of Interest: The authors declare no conflict of interest.

References

1. Majid, E.; Mahdi, G.; Mehdi, I. A review of zinc oxide mineral beneficiation using flotation method. *Adv. Colloid. Inter. Sci.* **2014**, *206*, 68–78. [CrossRef]
2. Hosking, N.C.; Ström, M.A.; Shipway, P.H.; Rudd, C.D. Corrosion resistance of zinc-magnesium coated steel. *Corros. Sci.* **2007**, *49*, 3669–3695. [CrossRef]
3. Zhao, H.B.; Gan, X.W.; Wang, J.; Tao, L.; Qiu, W.Q.; Qiu, G.Z. Stepwise bioleaching of Cu-Zn mixed ores with comprehensive utilization of silver-bearing solid waste through a new technique process. *Hydrometallurgy* **2017**, *171*, 374–386. [CrossRef]
4. Mudd, G.M.; Jowitt, S.M.; Werner, T.T. The world's lead-zinc mineral resources: Scarcity, data, issues and opportunities. *Ore Geol. Rev.* **2017**, *80*, 1160–1190. [CrossRef]
5. Irannajad, M.; Ejtemaei, M.; Gharabaghi, M. The effect of reagents on selective flotation of smithsonite–calcite–quartz. *Miner. Eng.* **2009**, *22*, 766–771. [CrossRef]
6. Moradi, S.; Monhemius, A.J. Mixed sulphide-oxide lead and zinc ores: Problems and solutions. *Miner. Eng.* **2011**, *24*, 1062–1076. [CrossRef]
7. Navidi Kashani, A.H.; Rashchi, F. Separation of oxidized zinc minerals from tailings: Influence of flotation reagents. *Miner. Eng.* **2008**, *21*, 967–972. [CrossRef]
8. Li, B.B.; Zhang, G.F.; Shi, Q. Atomic insights into the interaction mechanism underpinning carbonate ion adsorption in smithsonite flotation. *Process Saf. Environ. Prot.* **2024**, *190*, 11–21. [CrossRef]
9. Marabini, A.M.; Ciriachi, M.; Plescia, P.; Barbaro, M. Chelating reagents for flotation. *Miner. Eng.* **2007**, *20*, 1014–1025. [CrossRef]
10. Li, G.; Zhang, Z.; Wei, Q.; Jiao, F.; Yang, C.; Li, W.; Qin, W. Study on flotation recovery of typical carbon-bearing lead-zinc sulphide ore in Guizhou with pre-decarbonization. *Geochemistry* **2024**, *84*, 126096. [CrossRef]
11. Hosseini, S.H.; Forssberg, E. Smithsonite flotation using potassium amyl xanthate and hexylmercaptan. *Trans. Inst. Min. Metall. Sect. C Miner. Process. Extr. Metall.* **2006**, *115*, 107–112. [CrossRef]
12. Nanda, S.; Kumar, S.; Mandre, N.R. Flotation behavior of a complex lead-zinc ore using individual collectors and its blends for lead sulfide. *J. Dispers. Sci. Technol.* **2023**, *44*, 1703–1710. [CrossRef]
13. Wang, J.G.; Ji, Y.H.; Cheng, S.Y.; Liu, S.; Cao, J.; Chen, P. Selective flotation separation of galena from sphalerite via chelation collectors with different nitrogen functional groups. *Appl. Surf. Sci.* **2021**, *568*, 150956. [CrossRef]
14. Shen, Z.H.; Zhang, Q.; Fang, J. Research Progress in Surface Sulfidization of Smithsonite. *Nonferr. Met. Mieral Process. Sect.* **2021**, *01*, 37–46+59. [CrossRef]
15. Zhu, H.L.; Qin, W.Q.; Chen, C.; Chai, L.Y.; Li, L.S.; Liu, S.J.; Zhang, T. Selective flotation of smithsonite, quartz and calcite using alkyldiamineether as collector. *Trans. Nonferr. Metals Soc. China* **2018**, *28*, 163–168. [CrossRef]
16. Luo, B.; Nie, W.L.; Dong, J.S. Effect of Lead Ions on the Sulfidization Flotation of Smithsonite Using Sodium Butyl Xanthate as a Collector. *Miner. Eng.* **2022**, *185*, 107710. [CrossRef]
17. Ceylan, Y.; Altınışık, Y.; Cebeci, H.; Sis, L.K. A Process Mineralogy Approach to the Flotation of Complex Lead–Zinc Ores from Görgü (Malatya) Region. *Min. Met. Explor.* **2022**, *39*, 1219–1232. [CrossRef]
18. Mehdilo, A.; Zarei, H.; Irannajad, M.; Arjmandfar, H. Flotation of zinc oxide ores by cationic and mixed collectors. *Miner. Eng.* **2012**, *36*, 331–334. [CrossRef]
19. Feng, Q.C.; Wen, S.M. Formation of zinc sulfide specieson smithsonite surfaces and its response to flotation performance. *J. Alloys Compd.* **2017**, *709*, 602–608. [CrossRef]
20. Mehdilo, A.; Irannajad, M.; Zarei, H. Smithsonite Flotation from Zinc Oxide Ore using Alkyl Amine Acetate Collectors. *Sep. Sci. Technol.* **2014**, *49*, 445–457. [CrossRef]
21. Zhu, Y.H.; Huang, K.H.; Yang, S.Y.; Liang, Z.A. Studies of benzyl hydroxamic acid/calcium lignosulphonate addition order in the flotation separation of smithsonite from calcite. *Int. J. Min. Sci. Technol.* **2021**, *31*, 1153–1158. [CrossRef]
22. Barbery, G. Complex sulphide ores: Processing options. *Miner. Process. A Crossroads* **1986**, *117*, 157–194. [CrossRef]
23. Hosseini, S.H.; Forssberg, E. Adsorption studies of smithsonite flotation using dodecylamine and oleic acid. *Min. Metall. Explor.* **2006**, *23*, 87–96. [CrossRef]
24. Liu, C.; Feng, Q.M.; Zhang, G.F.; Chen, W.W.; Chen, Y.F. Effect of sodium carbonate on flotation behavior of smithsonite and its mechanism in presence of sodium oleate. *Chin. J. Nonferrous Met.* **2017**, *27*, 2379–2384. [CrossRef]
25. Ferreira, P.H.T.; Fernandes Lima, R.M. Concentration of oxidized Brazilian zinc ore by flotation: Comparative study between anionic and cationic routes. *Sep. Sci. Technol.* **2022**, *57*, 2625–2634. [CrossRef]

26. Sun, H.F.; Niu, F.S.; Zhang, J.X. Investigation on the flotation separation of smithsonite from calcite using calcium lignosulphonate as depressant. *Colloids Surf. A Physicochem. Eng. Asp.* **2021**, *630*, 127571. [CrossRef]
27. Shi, Q.; Feng, Q.M.; Zhang, G.F.; Deng, H. Electrokinetic properties of smithsonite and its floatability with anionic collector. *Colloids Surf. A Physicochem. Eng. Asp.* **2012**, *410*, 178–183. [CrossRef]
28. Hosseini, S.H.; Forssberg, E. Physicochemical studies of smithsonite flotation using mixed anionic/cationic collector. *Miner. Eng.* **2007**, *20*, 624. [CrossRef]
29. Hosseini, S.H.; Forssberg, E. Studies on selective flotation of smithsonite from silicate minerals using mercaptans and one stage desliming. *Miner. Process. Extr. Metall.* **2011**, *120*, 79–84. [CrossRef]
30. Majid, E.; Mehdi, I.; Mahdi, G. Influence of important factors on flotation of zinc oxide mineral using cationic, anionic and mixed (cationic/anionic) collectors. *Miner. Eng.* **2011**, *24*, 1402–1408. [CrossRef]
31. Wang, L.; Hu, G.Y.; Sun, W.; Khoso, S.A.; Liu, R.Q.; Zhang, X.F. Selective flotation of smithsonite from dolomite by using novel mixed collector system. *Trans. Nonferr. Met. Soc. China* **2019**, *29*, 1082–1089. [CrossRef]
32. Wang, Z.; Xu, L.H.; Wang, J.M.; Wang, L.; Xiao, J.H. A comparison study of adsorption of benzohydroxamic acid and amyl xanthate on smithsonite with dodecylamine as co-collector. *Appl. Surf. Sci.* **2017**, *426*, 1141–1147. [CrossRef]
33. Gao, Y.; Fu, X.Z.; Khoso, S.A.; Pan, Z.J.; Han, H.S.; Sun, W.; Yue, T. A quantitative innovation perspective on synergism and selectivity mechanism of mixed collectors in flotation. *Miner. Eng.* **2024**, *206*, 108474. [CrossRef]
34. Deng, R.R.; Wang, Y.; Duan, W.T.; Xing, D.Q.; Hu, Y. Induced crystallization of Pb^{2+} on smithsonite surface during sulfidation-xanthate flotation. *Colloids Surf.* **2022**, *650*, 129576. [CrossRef]
35. Wen, S.M. Thermodynamic theory of flotation for a complex multiphase solid–liquid system and high-entropy flotation. *Int. J. Min. Met. Mater.* **2024**, *31*, 1177–1197. [CrossRef]
36. Jia, K.; Feng, Q.M.; Zhang, G.F.; Ji, W.Y.; Zhang, W.K.; Yang, B.Q. The role of S(II) and Pb (II) in xanthate flotation of smithsonite: Surface properties and mechanism. *Appl. Surf. Sci.* **2018**, *442*, 92–100. [CrossRef]
37. Wang, M.T.; Zhang, G.F.; Chen, Y.F.; Zhao, L. Effect of surface oxidization on quartz slime coating in the sulfidization-amine flotation of smithsonite. *Miner. Eng.* **2022**, *188*, 107847. [CrossRef]
38. Cao, S.M.; Zhang, X.L.; Zeng, S.Q. The present situation and progress of zinc oxide flotation. *Adv. Mater. Res.* **2013**, *756*, 68–71. [CrossRef]
39. Liao, R.P.; Wen, S.M.; Liu, J.; Bai, S.J.; Feng, Q.C. Synergetic adsorption of dodecylamine and octyl hydroxamic acid on sulfidized smithsonite: Insights from experiments and molecular dynamics simulation. *Sep. Purif. Technol.* **2024**, *329*, 125106. [CrossRef]
40. Krystian, C.; Cezary, R.; Grazyna, P. Flotation of zinc and lead oxide minerals from Olkusz region calamine ores. *E3S Web Conf.* **2016**, *8*, 01042. [CrossRef]
41. Sun, Q.; Feng, Q.M.; Shi, Q. Adsorption mechanism of smithsonite by dodecyl phosphate ester potassium. *J. Cent. South Univ.* **2018**, *49*, 1845–1850. [CrossRef]
42. Arturo, B.T.; Roberto, P.G.; Diego, M.C. Zeta potential of air bubbles conditioned with typical froth flotation reagents. *Int. J. Min. Process* **2015**, *140*, 50–57. [CrossRef]
43. Wang, M.L.; Zhao, W.J.; Han, G.; Feng, Q.C. Utilization of lead ions to improve surface hydrophobicity and flotation recovery of sulfidized smithsonite. *Colloids Surf.* **2023**, *663*, 131126. [CrossRef]
44. Zhang, G.F.; Cui, M.M.; Zhu, Y.G.; Shi, Q.; Luo, N. Effect of water glass on flotation separation of smithsonite and quartz. *Chin. J. Nonferrous Met.* **2012**, *22*, 3535–3541.
45. Luo, B.; Liu, J.; Liu, Q.; Song, C.; Yu, L.; Li, S.; Lai, H. A Mechanism for the Adsorption of 2-(Hexadecanoylamino)Acetic Acid by Smithsonite: Surface Spectroscopy and Microflotation Experiments. *Minerals* **2019**, *9*, 15. [CrossRef]
46. Zeng, Y.; Liu, J.; Dong, W.C.; Hao, J.M.; Wang, Y. Study on Sulfide Layer Attenuation Behavior of Smithsonite During Sulfidization Flotation. *Front. Mater.* **2019**, *6*, 347. [CrossRef]
47. Zeng, Y.H.; Yao, X.; Liu, G.X.; He, G.M.; Yu, X.Y.; He, G.C.; Huang, Z.Q.; Zhang, R.R.; Cheng, C. Flotation behavior and mechanism of phenylpropenyl hydroxamic acid for the separation of smithsonite and calcite. *J. Mol. Liq.* **2021**, *339*, 116893. [CrossRef]
48. Zhao, L.; Liu, W.G.; Liu, W.B.; Zhou, S.J.; Peng, X.Y. Investigation on matching relationship between surface characters and collector properties: Achieving flotation separation of zinc oxide minerals from quartz. *Colloids Surf. A* **2021**, *617*, 126392. [CrossRef]

minerals

MDPI

Article

Investigation of a Novel Depressant for Flotation Separation of Chalcopyrite and Galena: Experiments and Adsorption Mechanisms

Hong Zeng [1,2], Chongjun Liu [2,*], Tong Lu [2], Zehui Gao [2], Yangge Zhu [2,*], Chuanyao Sun [2], Zhiqiang Zhao [2], Guiye Wu [2], Ruidong Li [3] and Jun Hu [3]

[1] School of Civil and Resources Engineering, University of Science and Technology Bejing, Bejing 100083, China; zenghong@bgrimm.com

[2] State Key Laboratory of Mineral Processing, BGRIMM Technology Group, Beijing 102600, China; lutong@bgrimm.com (T.L.); gzh17702971821@163.com (Z.G.); suncy@cae.com (C.S.); zhaozhiqiang@bgrimm.com (Z.Z.); wuguiye@bgrimm.com (G.W.)

[3] School of Chemical Engineering, Northwest University, Xi'an 710069, China; liruidong20231004@163.com (R.L.); hujun@nwu.edu.cn (J.H.)

* Correspondence: liuchongjun@bgrimm.com (C.L.); zhuyangge@126.com (Y.Z.)

Abstract: To reduce lead content in copper concentrates, this study developed a novel galena depressant, TA (thioureidoacetic acid). This study utilizes a synthetic mineral feed with fully liberated galena and chalcopyrite from separate sources to establish baseline separation conditions. The adsorption capability of TA on galena surfaces was systematically investigated through micro-flotation tests, surface characterization, and first-principles calculations. Results demonstrate that TA effectively reduces galena recovery (from 82.92% to 12.29%) without compromising chalcopyrite flotation efficiency (>83.2% recovery) when using thionocarbamate (Z200) as the collector. FTIR and XPS analyses confirm that TA chemisorbs onto galena surfaces via its C=S and C=O functional groups. First-principles calculations reveal dual Pb-S and Pb-O bond formation during TA adsorption, resulting in stronger interfacial binding energy compared to Z200. This work establishes a molecular engineering framework for designing high-selectivity depressants.

Keywords: DFT; chalcopyrite; galena; depressant

1. Introduction

Copper, as a cornerstone material of modern industrialization, continues to drive global demand growth due to its irreplaceable role in power infrastructure, renewable energy systems, and advanced manufacturing [1]. Chalcopyrite serves as the primary source of copper resources and is naturally associated with galena in mineral deposits. During flotation processes, while the industrial standard collector Z200 (O-isopropyl-N-ethyl thiocarbamate) significantly enhances chalcopyrite recovery by inducing surface hydrophobicity, its non-selective adsorption mechanism concurrently activates galena surfaces, resulting in persistently elevated lead impurity levels [2]. These lead impurities not only diminish the economic value of copper products but also substantially escalate smelting costs due to toxic lead emissions during pyrometallurgical processes [3,4]. Consequently, there is an urgent demand—both economically and environmentally—for highly efficient flotation depressants capable of reducing lead content in copper concentrates during beneficiation.

Current galena depressant research primarily focuses on inorganic and organic inhibitor systems with distinct environmental and efficiency characteristics. The commonly used depressants for galena are listed in Table 1. Traditional inorganic depressants represented by potassium dichromate ($K_2Cr_2O_7$) have demonstrated effective galena depression through surface oxidation mechanisms, achieving copper–lead separation efficiency through $PbCrO_4$ film formation on galena surfaces. However, hexavalent chromium's carcinogenicity severely limits its industrial application [5]. Sulfide-based systems like sodium sulfide (Na_2S) also face environmental restrictions due to potential acid mine drainage generation [6–9].

Table 1. Galena depressant reagents in use.

Inorganic Reagents	Organic Macromolecule	Small Organic Molecule
Potassium dichromate	Polyacrylic acids	5-amino-1,3,4-thiadiazole-2-thiol
Sodium sulfide	Polyacrylamide	1,3-oxathiolane derivatives
Fe^{2+}/NaClO	Modified lignosulfonates	N-(phosphonomethyl) iminodiacetic acid
Sodium periodate	Phosphonyl carboxylic acid copolymer	DL-dithiothreitol
	Pinewood pyrolysis liquid	
	Dextrin	
	Tannic acid	

Organic depressant research has evolved into three principal directions: polymeric reagents, plant-derived compounds, and small-molecule systems. Polyacrylic acid derivatives such as polyacrylamide (PAAS) combined with H_2O_2 oxidants show selective depression through chelation with Pb^{2+} ions and surface hydroxylation [10–12], achieving comparable separation efficiency to chromium reagents while maintaining 83.5% Cu recovery [5]. Similarly, phosphonyl carboxylic acid copolymer (POCA) demonstrates selective adsorption on galena through Pb-O-P bonding interactions [13]. However, these polymeric depressants frequently exhibit cross-mineral suppression effects, particularly reducing chalcopyrite recoveries at elevated dosages.

Plant-derived depressants, including dextrin and tannins, show promising environmental compatibility [14–17]; however, their performance is pH-dependent. Modified lignosulfonates exhibit strong depressive effects on galena during flotation; however, their lack of selectivity also suppresses chalcopyrite recovery [18]. Recent studies reveal that pinewood pyrolysis liquid (PL) achieves >90% Pb recovery with <5% pyrite recovery through phenolic compound adsorption but requires precise pH control (pH 8.5–9.0) for optimal performance [19]. Small-molecule systems present new opportunities through targeted molecular design. Compounds like N-(phosphonomethyl) iminodiacetic acid (PMIDA) inhibit sphalerite via hydrophilic Zn-O-P complexation while maintaining galena floatability, attaining a separation efficiency of 84.3% [20]. The Fe^{2+}/NaClO system generates ·OH radicals to oxidize PbS surfaces into hydrophilic $PbSO_4$ coatings, achieving 9.48% PbS recovery versus 82.33% $CuFeS_2$ recovery [21]. 5-amino-1,3,4-thiadiazole-2-thiol (MATT) and 1,3-oxathiolane derivatives have shown promising results, yet their complex synthesis routes and high raw material costs hinder industrial scalability [22,23].

Despite these advancements, key challenges persist: (1) Orbital interaction mechanisms between depressant functional groups and galena surfaces remain poorly understood, particularly regarding d-orbital electron transfer in Pb-S-depressant complexes; (2) Most organic depressants require combined oxidants (e.g., H_2O_2) or activators, complicating reagent regimes; (3) Industrial validation of novel reagents like sodium periodate (SP) and DL-dithiothreitol (DLD) remains limited [24,25].

Prior studies have largely focused on empirical observations of adsorption behavior, leaving the electronic origins of selectivity underexplored. A critical gap persists in understanding the orbital interaction mechanisms governing depressant adsorption on galena surfaces. To address this, we designed thioureidoacetic acid (TA), a novel depressant featuring dual functional groups (C=S and C=O) that synergistically chemisorb onto galena via Pb-S and Pb-O bonds. By integrating COHP analysis with DFT calculations, particularly through crystal orbital Hamiltonian population (COHP) analysis—an essential tool for quantifying bond strength and electronic coupling—this work deciphers the orbital-level interactions that underpin TA's enhanced adsorption energy, thereby establishing a new design paradigm for flotation depressants rooted in orbital engineering. In this study, we focus on a pure mineral system where galena and chalcopyrite are fully liberated and sourced independently. This approach allows us to isolate key variables under idealized conditions before addressing complex naturally occurring ores in future work. Micro-flotation and synthetic mixed-mineral tests demonstrate that TA selectively inhibits lead–copper minerals in Z200-containing pulp systems. Fourier transform infrared spectroscopy (FTIR) and X-ray photoelectron spectroscopy (XPS) analyses reveal that TA achieves specificity through electron-transfer-mediated chemisorption on galena surfaces. This work establishes a new paradigm for eco-efficient chalcopyrite–galena separation and provides an innovative solution for sustainable mineral resource utilization, significantly advancing environmentally friendly practices in mineral processing.

2. Materials and Methods

2.1. Minerals and Reagents

High-grade galena and chalcopyrite samples were obtained from a mine in Hubei Province, China. The samples underwent crushing and fine grinding using a ceramic ball mill. Subsequently, particle sizes ranging from −74 μm to +38 μm were dry-screened for micro-flotation experiments. Figure 1 illustrates the X-ray diffraction (XRD) spectra of chalcopyrite and galena, while Table 2 presents the results of multi-element analysis for both minerals. Based on XRD and multi-element analysis, it is evident that the galena and chalcopyrite mineral samples exhibit exceptional purity levels, with galena purity reaching 98.14% and chalcopyrite purity at 98.23%. Therefore, the galena sample boasts a purity level of 98.14%, whereas the chalcopyrite sample demonstrates a purity level of 98.23%.

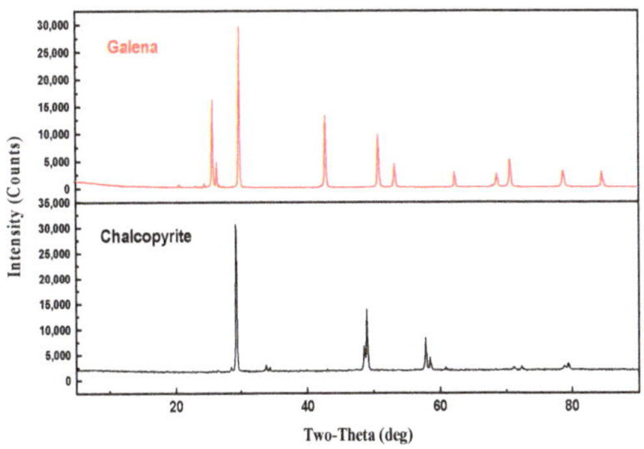

Figure 1. X-ray diffraction (XRD) results of the galena and chalcopyrite samples.

Table 2. Multi-element assay of the galena and chalcopyrite samples.

Sample	P	Zn	Pb	SiO_2	Al_2O_3	CaO	MgO	Cu	Tfe	S
Chalcopyrite	0.13	0.075	0.016	<0.01	0.040	0.11	0.078	33.95	30.84	34.54
Galena	0.0021	<0.01	84.99	<0.01	0.047	0.032	0.013	0.031	0.067	12.11

The Z200 utilized in the flotation experiment as a collector is sourced from BGRIMM Chemical Technology Co., Ltd., Cangzhou, China. The frother (MIBC) was procured through McLean Biochemical Technology Co., Ltd., Shanghai, China. Sodium hydroxide (NaOH) and hydrochloric acid (HCl), employed as pH adjusters, were acquired from Sinopharm Chemical reagent Co., Ltd., Shanghai, China.

The synthetic route of TA is shown in Figure 2. The required product is obtained by adding chloroacetic acid and NaOH to thiourea and heating the reaction for 5 h. The NMR figure of TA is shown in Figure 3. 13C-NMR (150 MHz, Deuterium Oxide): δC 201.6, 179.5, 178.6, 58.5, 39.2. 1H-NMR (600 MHz, Deuterium Oxide): 4.70 (d, J = 1.4 Hz, 1H), 3.89 (s, 1H), 3.19 (s, 1H).

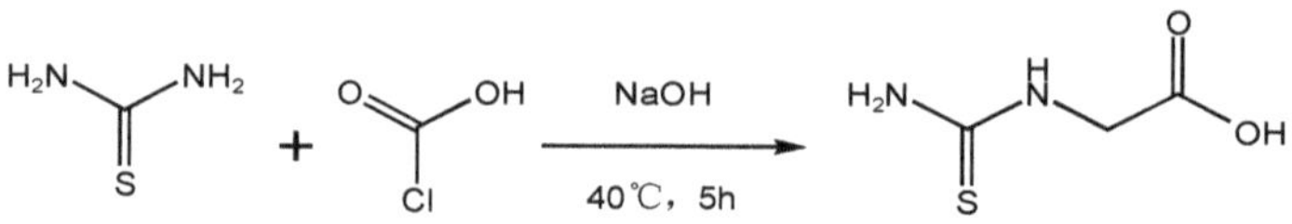

Figure 2. Synthesis route of TA.

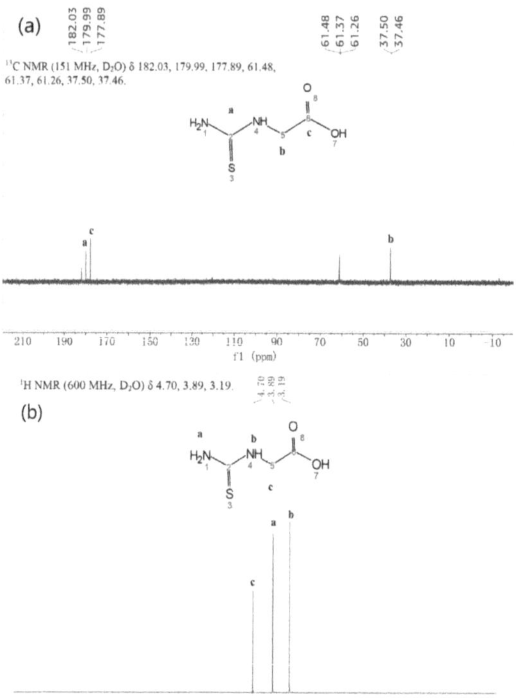

Figure 3. 13C NMR (**a**) and 1H NMR (**b**) spectra of TA.

2.2. Methods

2.2.1. Flotation Experiments

Pure mineral micro-flotation experiments were performed using a 30 mL XFGII flotation cell, operating at 1758 rpm without the need for an external air supply. Before each pure mineral flotation test, 2.0 g of pure mineral sample was placed in a beaker containing 30 mL of deionized water and ultrasonically cleaned for 5 min to remove oxides from the mineral surface. After 10 min of settling, the suspension was decanted, and the bottom pure minerals were transferred to a flotation cell. Then, the required flotation reagents, such as NaOH, HCl, TA, Z-200, and MIBC, were introduced into the slurry one by one during the stirring process. The slurry was stirred for 2 min after each addition of reagent. After the introduction of all reagents, the flotation process proceeded for 4 min. Finally, the foam product and residue were filtered, dried, and weighed. Each experiment was repeated three times to obtain the average yield. The flotation process diagram is presented in Figure 4.

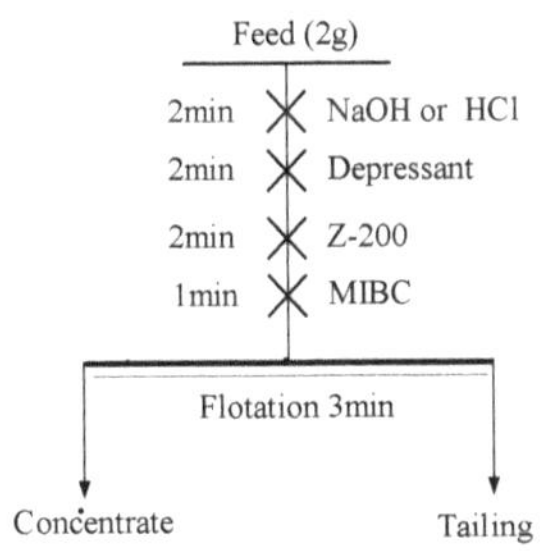

Figure 4. Micro-flotation procedure.

All micro-flotation experiments were performed in triplicate (n = 3), and results are expressed as mean ± standard deviation (SD). Statistical analyses, including one-way analysis of variance (ANOVA), were conducted manually using critical values from standard statistical tables to ensure rigor and reproducibility. The significance level was set at $\alpha = 0.05$.

2.2.2. FTIR Measurements

The adsorption behavior of TA on galena surfaces was investigated through Fourier transform infrared spectroscopy (FTIR) by analyzing TA, pristine minerals, and TA-modified specimens. The experimental workflow consisted of sequential collector preconditioning, depressant treatment, and spectroscopic characterization. For preconditioning, precisely measured 2.0 g aliquots of galena were introduced into 50 mL glass reactors containing 30 mL of 10 mg/L TA solution. Homogenization was achieved via mechanical agitation (300 rpm, 5 min) to facilitate complete collector adsorption, followed by vacuum filtration and thermal stabilization at 50 °C until mass equilibrium. Post-reaction solids were isolated through Buchner filtration and vacuum desiccated to constant mass, yielding TA-adsorbed mineral surfaces. Infrared spectral acquisition involved pellet preparation by homogenizing reference materials (TA) and mineral samples (pristine/treated) with anhydrous KBr (1:100 mass ratio) under 10 MPa compaction. Spectral data spanning 500–4000 cm^{-1} were collected using a Nicolet IS10 spectrometer (Thermo Fisher Scientific, Waltham, MA, USA) configured with 32 cumulative scans at 4 cm^{-1} spectral resolution.

2.2.3. XPS Measurements

Surface chemical evolution during TA adsorption was systematically probed through X-ray photoelectron spectroscopy (XPS) by analyzing pristine galena and TA-modified galena. Surface-modified specimens were prepared following identical pretreatment protocols as those employed for FTIR sample fabrication. XPS characterization was executed using a Thermo Scientific K-Alpha spectrometer equipped with a monochromatic Al Kα X-ray source (1486.6 eV). Prior to analysis, the ultrahigh vacuum chamber was evacuated to a base pressure of 1×10^{-9} Torr to minimize surface contamination. Wide-scan spectra spanning 0–1200 eV binding energy were acquired with a pass energy of 150 eV and step size of 1.0 eV. To ensure statistical reliability, triplicate measurements were performed for each specimen under identical analytical conditions. Charge compensation was achieved using a dual-beam electron flood gun, and energy scale calibration was rigorously implemented through reference to adventitious carbon (C 1s at 284.8 eV) [26–28].

2.2.4. Computation Approach

To elucidate the competitive adsorption mechanisms between collector Z200 and depressant TA on galena surfaces, density functional theory (DFT) calculations were performed using the Vienna Ab initio Simulation Package (VASP) to investigate adsorption configurations and electron transfer behaviors [28,29]. The generalized gradient approximation (GGA) with the Perdew–Burke–Ernzerhof (PBE) functional was employed to simulate electron density gradients and interfacial interactions [30], ensuring quantum-mechanically rigorous modeling of electronic properties. Computational parameters were optimized for precision: a 450 eV energy cutoff balanced computational efficiency and accuracy, while a $4 \times 2 \times 1$ Monkhorst–Pack k-point grid ensured adequate Brillouin zone sampling. Convergence criteria were strictly enforced (energy $<1 \times 10^{-5}$ eV, forces <0.02 eV/Å) to guarantee reliable electronic structure solutions. The configurations of valence electrons of Pb, C, N, S, H, and O were performed as $6s^26p^2$, $2s^22p^2$, $2s^22p^3$, $3s^23p^4$, $1s^1$, and $2s^22p^4$, respectively. The Projector Augmented Wave (PAW) method accurately modeled core–valence electron interactions, complemented by the DFT-D3 dispersion correction for long-range van der Waals forces [31–34]. Structural visualizations and electronic analyses were conducted using the VESTA 3.9 software package [35].

The galena crystal structure, sourced from established crystallographic databases [36,37], was geometrically optimized and cleaved along the (100) surface—the predominant cleavage plane under mechanical stress [38,39]. A four-layer slab model with a 20 Å vacuum buffer was constructed to minimize periodic boundary artifacts. Final DFT simulations utilized a ($4 \times 2 \times 1$) supercell configuration.

The adsorption capacity of flotation reagents on the mineral surface is evaluated based on their adsorption energy, which can be determined using Equation (1).

$$E_{ads} = E_{adsorbates/surface} - E_{surface} - E_{adsorbates} \quad (1)$$

Here, E_{ads} is the adsorption energy, $E_{adsorbates/surface}$ is the computed energy of the flotation reagents adsorbed on the galena/chalcopyrite surface, $E_{surface}$ is the energy of the galena/chalcopyrite surface, and $E_{adsorbates}$ is the energy of the flotation reagents.

3. Results

3.1. Micro-Flotation Experiments

Figure 5a illustrates the influence of TA dosage (0–60 mg/L) on the flotation recovery of chalcopyrite and galena at pH 10. As evidenced by the data, chalcopyrite recovery remains consistently above 92.3% (SD = 1.5%, 95% CI: 90.8–93.8%) across all tested TA concentrations, demonstrating negligible sensitivity to TA presence. In contrast, galena exhibits a stark dependency: its recovery decreases progressively with increasing TA concentration. In the absence of TA, galena recovery reaches 91.88% (SD = 1.2%, 95% CI: 90.5–93.3%), comparable to chalcopyrite. At the critical TA concentration of 60 mg/L, galena recovery plummets to 7.5% (SD = 0.8%, 95% CI: 6.9–8.1%; one-way ANOVA, $p < 0.001$), confirming TA's potent selective depression effect on galena while preserving chalcopyrite floatability.

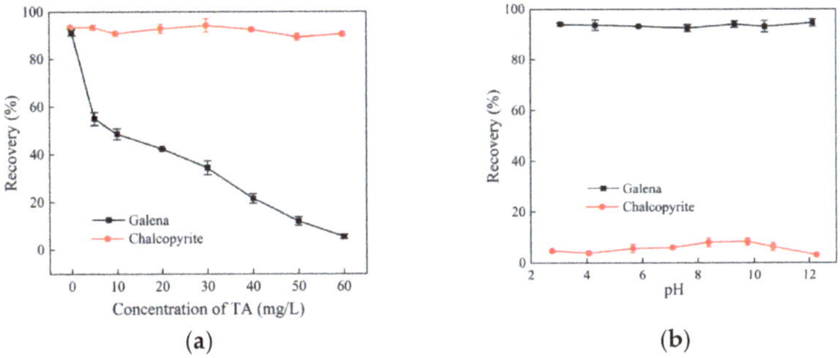

Figure 5. (**a**) Impact of TA dosage on the flotation recovery rates of chalcopyrite and galena; (**b**) influence of pH on the flotation recovery of chalcopyrite and galena (TA = 60 mg/L).

Figure 5b demonstrates the pH-dependent flotation behavior (pH 2–13) under optimized TA concentration (60 mg/L). Chalcopyrite maintains exceptional recovery stability (>93.4%, SD = 1.2%, 95% CI: 92.2–94.6%) across the entire pH spectrum, unaffected by TA addition. Conversely, galena recovery remains suppressed to approximately 8.5% (SD = 0.9%, 95% CI: 7.6–9.4%; ANOVA, $p < 0.001$) regardless of pH variations. This pH-independent suppression mechanism establishes a robust operational window for selective separation, with the maximum differential in recovery between chalcopyrite and galena reaching 80.2% (SD = 1.7%, 95% CI: 78.5–82.0%). TA maintains stable suppression across pH 2–13 (Figure 5b); however, this behavior was validated under controlled laboratory conditions. Real-world applications may require additional optimization for ore matrices containing magnesian gangue minerals, which could alter solution chemistry and reagent efficacy.

3.2. *Artificially Mixed Minerals Experiments*

To further investigate the selective depression capability of TA on galena during flotation, artificial mixed-mineral flotation separation tests of chalcopyrite and galena were conducted in this study. The experimental results are presented in Figure 6. As shown in Figure 6a, when the TA concentration in the pulp increased from 10 mg/L to 50 mg/L, the flotation recovery of galena in the concentrate significantly decreased from 82.92% (SD = 2.1%, 95% CI: 80.8–85.0%) to 12.29% (SD = 1.5%, 95% CI: 10.8–13.8%; one-way ANOVA, $p < 0.001$), accompanied by a reduction in the galena grade from 51.27% (SD = 1.8%, 95% CI: 49.5–53.1%) to 13.27% (SD = 1.2%, 95% CI: 12.1–14.5%, one-way ANOVA, $p < 0.001$). Notably, further increases in TA concentration beyond 50 mg/L had minimal impact on the grade and recovery of galena in the concentrate. Conversely, the recovery of copper remained relatively stable at approximately 83.2% (SD = 1.8%, 95% CI: 81.4–85.0%), demonstrating negligible sensitivity to TA concentration.

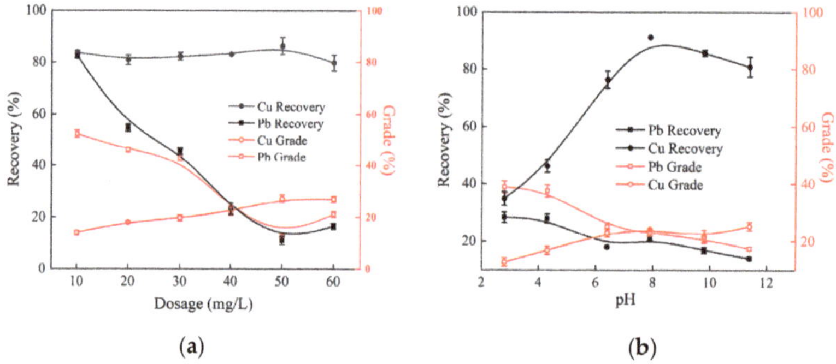

Figure 6. (**a**) Impact of TA dosage on the recovery of a chalcopyrite–galena mixed mineral; (**b**) influence of pH on the recovery of the chalcopyrite–galena mixed mineral (TA = 50 mg/L).

Figure 6b illustrates the influence of pulp pH (2.8–11.4) on the flotation performance of chalcopyrite and galena at a fixed TA concentration of 50 mg/L. As the pH increased from 2.8 to 11.4, the recovery of galena in the concentrate decreased from 29.65% (SD = 2.3%, 95% CI: 27.4–31.9%) to 13.78% (SD = 1.6%, 95% CI: 12.2–15.4%; ANOVA, $p < 0.001$), while its grade declined from 40.32% (SD = 2.0%, 95% CI: 38.3–42.3%) to 20.03% (SD = 1.4%, 95% CI: 18.6–21.5%). In contrast, the recovery of copper initially increased from 36.48% (SD = 2.5%, 95% CI: 34.0–39.0%) to 91.21% (SD = 1.7%, 95% CI: 89.5–92.9%, ANOVA, $p < 0.001$) as the pH rose from 2.8 to 7.9, followed by a slight reduction to 83.28% (SD = 1.9%, 95% CI: 81.4–85.2%) at pH 11.4. These results confirm that TA exhibits a pronounced selective depression effect on galena flotation across a wide pH range.

3.3. FTIR Experiments

This study analyzed the adsorption behavior of TA on the surface of galena. FTIR was used to examine galena particles treated with these agents. Figure 7 shows the FTIR spectra of TA, galena, and TA-treated galena to analyze the impact of TA on galena. Figure 6 reveals that TA exhibits characteristic FTIR absorption peaks at 1652.91 cm^{-1} (C=O stretching vibration), 1574.67 cm^{-1} (N-H bending vibration), and 1396.54 cm^{-1}, 1074.11 cm^{-1}, 1228.02 cm^{-1} corresponding to C=S stretching vibration, -OH stretching vibration, and C-O stretching vibration, respectively. Comparing the infrared spectra of galena and TA-treated galena, the TA-treated galena shows modified absorption peaks at 1633.11 cm^{-1} (C=O stretching vibration) and 1361.76 cm^{-1} (C=S stretching vibration). The observed bathochromic shifts in characteristic vibrational peaks ($\Delta\nu$: C=O = -19.8 cm^{-1}, C=S = -34.78 cm^{-1}) relative to pure TA spectra demonstrate that TA predominantly chemisorbs onto galena surfaces through covalent interactions involving its C=S and C=O functional groups [40–43]. This chemical anchoring mechanism, evidenced by the formation of Pb-S and Pb-O coordination bonds, fundamentally differs from physical adsorption processes and confirms the specific electronic interactions between TA's electron-donating moieties and surface lead atoms.

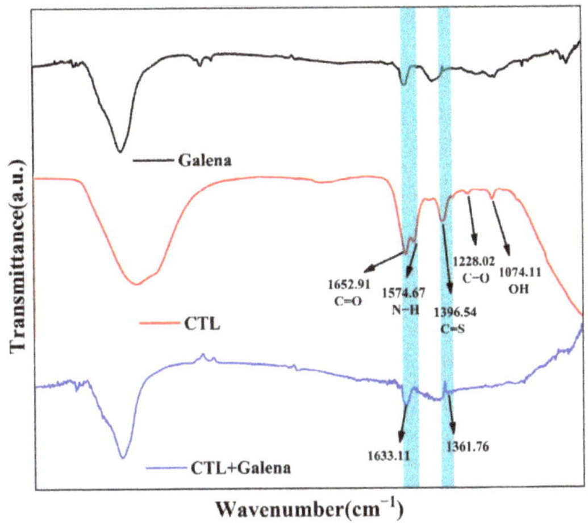

Figure 7. FTIR spectra of TA, untreated and treated galena.

3.4. XPS Experiments

The analytical results presented in Figure 8 show the XPS measurements of galena surfaces before and after TA treatment. As clearly demonstrated in the figure, the nitrogen content on the galena surface increased significantly from 0.66% to 5.47% following TA treatment, which indicates substantial adsorption of TA molecules onto the mineral surface after the modification process.

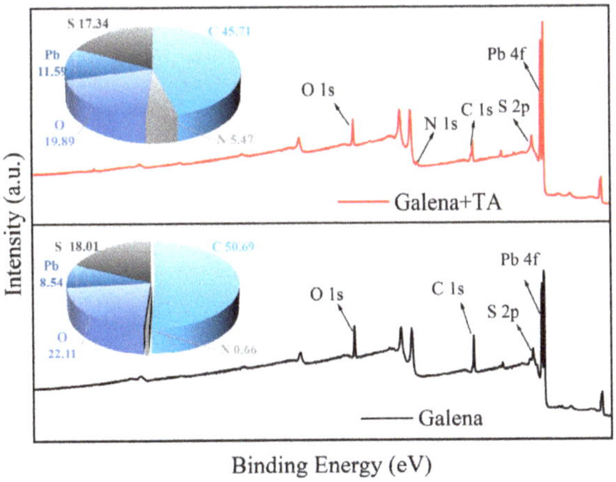

Figure 8. Comparative XPS wide-scan analysis of galena surfaces before and after treatment.

To further clarify the electron transfer during the adsorption process of TA on the surface of galena, the characteristics of S 2p, C 1s, Pb 4f, and O 1s were analyzed in detail through high-resolution XPS spectroscopy.

Figure 9a shows the high-resolution spectra of the S element on the surface of galena before and after treatment with TA. It can be seen from the figure that before the treatment with TA, the S element on the surface of galena mainly has two peaks at 160.70 and 161.90 eV [41], which are the doublet peaks from S 2p3/2. The doublet peaks at 168.10 and 167.40 eV are from the sulfates formed by the oxidation of the S element. After the treatment with TA, the two peaks at 160.70 and 161.90 eV on the surface of galena remain unchanged, indicating that the S on the surface of galena does not participate in the reaction during the adsorption of TA. After TA is adsorbed on the surface of galena, a new peak at 163.70 eV appears in the high-resolution spectrum of the S element on the surface, which is from the S in TA. In Figure 9b, after the treatment with TA, a new absorption peak of the C element from TA appears at 287.50 eV, indicating that TA is adsorbed on the surface of galena after the treatment.

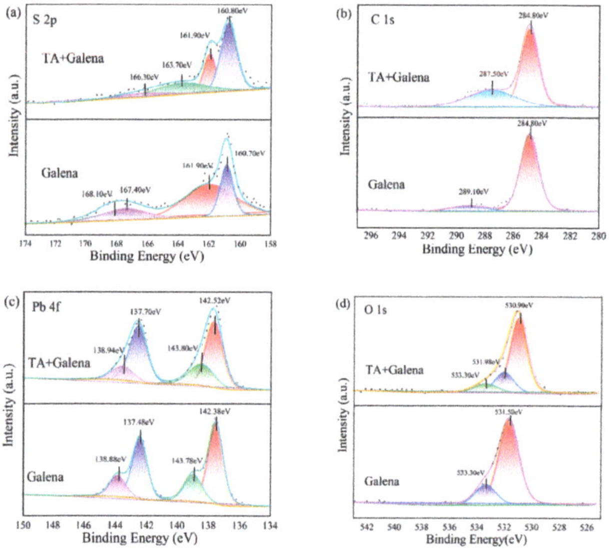

Figure 9. XPS spectra of the (**a**) S 2p, (**b**) C 1s, (**c**) Pb 4f, and (**d**) O 1s on the surface of untreated and treated galena.

Figure 9c reflects the high-resolution spectra of Pb 4f on the surface of galena before and after treatment with TA. It can be seen from the figure that before the treatment with TA, the high-resolution spectrum of Pb 4f mainly has characteristic peaks at 137.48 and 142.38 eV. After the treatment with TA, the characteristic peaks change to 137.70 and 142.52 eV. The decrease in the binding energy indicates that during the adsorption of TA on the surface of galena, there is an electron transfer from TA to the Pb on the surface of galena [42].

Figure 9d shows the high-resolution spectra of O 1s on the surface of galena before and after treatment with TA. It can be seen from the figure that compared with the surface of galena before the treatment with TA, the surface of galena after the treatment with TA has an additional characteristic peak at 531.98 eV for O 1s, which is from the -COOH group in TA, indicating that TA is adsorbed on the surface of galena.

3.5. Adsorption Mechanisms from DFT Calculation

Figure 10a depicts the adsorption configuration of the collector Z200 on the surface of galena. It can be observed from the figure that Z200 adsorbs onto the surface of galena through the S atom in its molecule. The bond length between the S atom and the Pb atom is 3.11 Å, which is less than the sum of the atomic radii of Pb and S (4.10 Å). This indicates that the adsorption of Z200 on the surface of galena is a chemical adsorption. Figure 10b shows the differential charge density map of the adsorption of the collector Z200 on the surface of galena. These maps display electron depletion (blue) and accumulation (yellow). Notably, electron depletion can be clearly observed around the S atom in Z200, while electron accumulation occurs around the Pb atom on the surface of galena. This suggests that the electrons in the flotation reagent are transferred to the surface of galena.

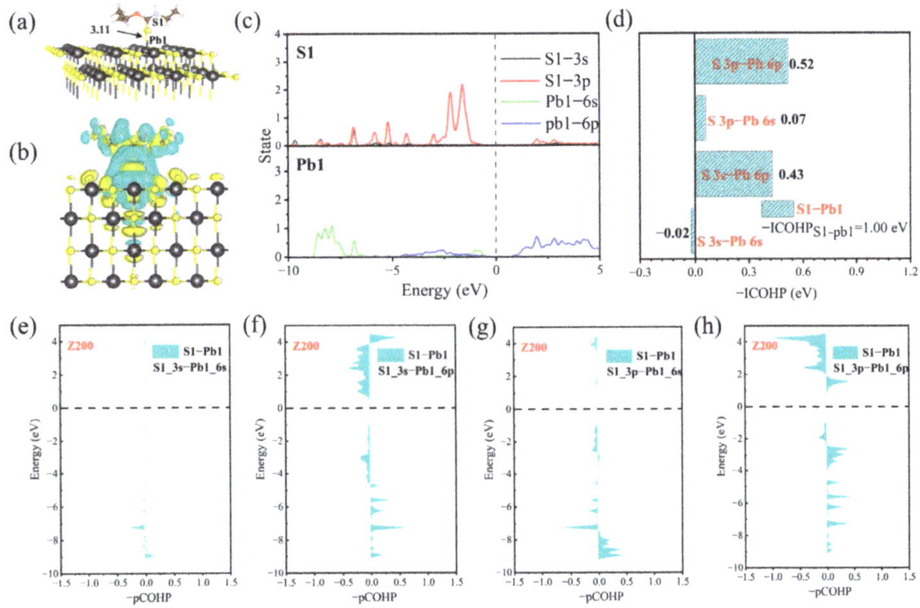

Figure 10. Electronic and structural features of Z200-galena interactions: (**a**) adsorption patterns, (**b**) differential charge distribution, (**c**) density of states projection, (**d**) bond strength integration (-ICOHP), (**e**–**h**) orbital interaction analysis (COHP).

To elucidate the electronic interactions between the flotation reagent and lead species on galena surfaces, the projected density of states (PDOS) was employed to characterize sulfur (Z200) and surface-bound lead (galena). Figure 10c demonstrates distinct orbital distributions: the S 3s orbital predominantly occupies the low-energy regime (<−5 eV), whereas the S 3p orbital spans the higher-energy range (−5 to 0 eV). In contrast, Pb 6s orbitals localize within −10 to −5 eV, while Pb 6p orbitals overlap with the S 3p orbitals in the −5 to 0 eV region. This spatial-energy alignment suggests complementary hybridization between S 3p and Pb 6p orbitals, while Pb 6s orbitals remain energetically isolated.

To understand how Z200 binds to galena, we analyzed the electronic interactions using a method called COHP [43,44]. This revealed that the strongest bond forms between sulfur's 3p orbital and lead's 6p orbital, contributing 0.52 eV to the total bond strength of 1.00 eV. A weaker interaction (0.43 eV) occurs between sulfur's 3s and lead's 6p orbitals, while lead's 6s orbitals play almost no role. Essentially, Z200's sulfur atom "shares" electrons primarily with lead's 6p orbitals, creating a stable bond that anchors it to the mineral surface; however, this single-bond mechanism limits its overall effectiveness [45,46].

The strength of these bonds directly impacts Z200's ability to suppress galena during flotation. While its 1.00 eV bond is strong enough to reduce galena recovery, TA outperforms Z200 by forming two bonds (S–Pb and O–Pb) with a combined strength of 1.76 eV. This dual-bond strategy not only enhances adsorption stability but also explains TA's superior selectivity and pH resilience, offering a blueprint for designing next-generation depressants that prioritize multi-site bonding for industrial applications.

Figures 11a and 12a present the adsorption configurations of TA on the surface of galena. It can be seen from the figures that the S atom in the TA molecule forms an S2-Pb2 bond with the Pb atom on the surface of galena, and the O atom in the TA molecule forms an O1-Pb3 bond with the Pb atom on the surface of galena. Similar to the collector, the bond length of the S2-Pb2 bond is 3.11 Å, indicating that a chemical bond is also formed between S2 and Pb2. The bond length between O1 and Pb3 is 2.84 Å, which is less than the sum of the atomic radii of O and Pb (3.54 Å), suggesting that a chemical bond is formed between the oxygen atom and the lead atom as well.

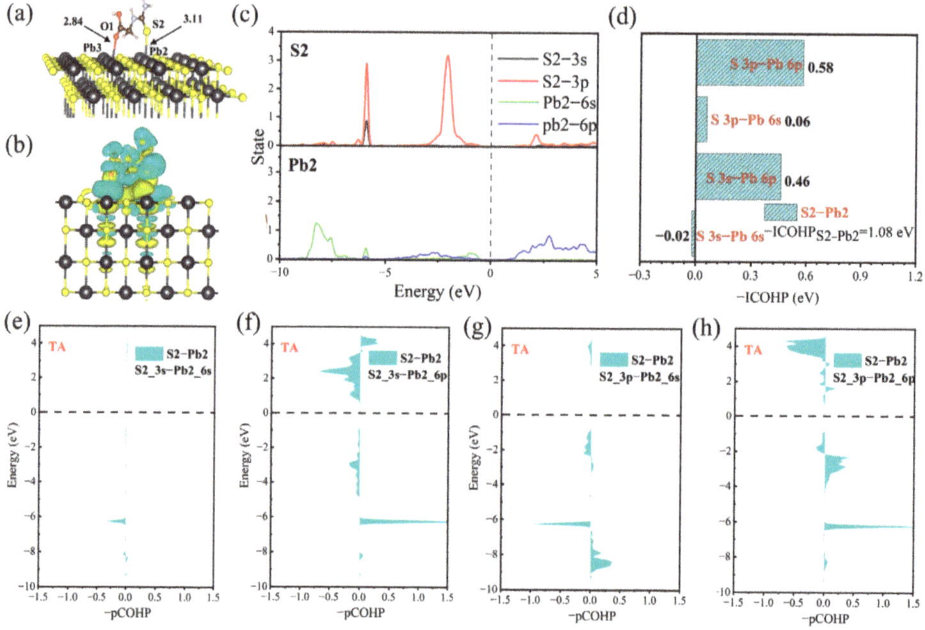

Figure 11. Adsorption configurations (**a**), differential charge density (**b**) of TA on the galena surface and PDOS (**c**), -ICOHP (**d**), and COHP (**e–h**) analysis of S-Pb bonds formed by TA adsorption on the surface of galena.

Figures 11b and 12b show the differential charge density maps of TA adsorbed on the surface of galena. It can be clearly observed that electron depletion occurs around the S2 and O1 atoms in the TA molecule, and electron accumulation is found around the Pb2 and Pb3 atoms on the surface of galena. This indicates that when TA is adsorbed on the surface of galena, electrons are transferred from TA to the surface of galena. This electron transfer is consistent with the results of XPS.

The projected density of states (PDOS) of S in TA and Pb on the surface of galena was analyzed. As shown in Figure 11c, the S 3s orbital is mainly located in the low-energy region below −5 eV, while the S 3p orbital has two sharp peaks in the range from −7.5 eV to 0 eV. In contrast, the Pb 6s orbital is mainly located in the low-energy spectrum from −10 eV to −5 eV, and the Pb 6p orbital is also in the high-energy region from −5 eV to 0 eV. This indicates that the S 3p and Pb 6p orbitals have similar high-energy distributions, while the Pb 6s orbital is significantly concentrated in the low-energy region.

The projected density of states (PDOS) of O in TA and Pb on the surface of galena was analyzed. As can be seen from Figure 12c, the O 2s and O 2p orbitals are mainly located in the low-energy region below −5 eV. The Pb 6s orbital is mainly located in the low-energy spectrum from −10 eV to −5 eV, and the Pb 6p orbital is in the high-energy region from −5 eV to 0 eV.

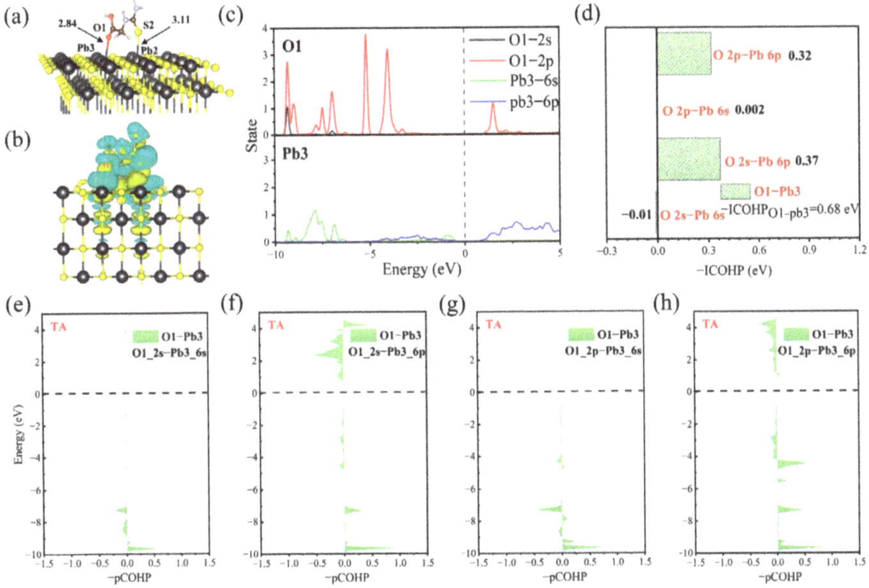

Figure 12. Adsorption configurations (**a**), differential charge density (**b**) of TA on the galena surface and PDOS (**c**), -ICOHP (**d**), and COHP (**e**–**h**) analysis of O-Pb bonds formed by TA adsorption on the surface of galena.

The S2-Pb2 bond was analyzed by COHP. The -ICOHP value of the S2-Pb2 bond formed by the adsorption of TA on the surface of galena is 1.08 eV, which is similar to that of the S1-Pb1 bond formed by the adsorption of the collector on the mineral surface. It can be seen from Figure 11d–h that the interaction strength between the S2 and Pb2 orbitals is similar to that between the S1 and Pb1 orbitals. The interactions between the S 3s and Pb 6s orbitals and between the S 3p and Pb 6s orbitals contribute little to the bonding, while the interactions between the S 3s and Pb 6p orbitals and between the S 3p and Pb 6p orbitals contribute significantly to the bonding, with values of 0.46 eV and 0.58 eV, respectively.

The O1-Pb3 bond was analyzed by COHP. The -ICOHP value of the S2-Pb2 bond formed by the adsorption of TA on the surface of galena is 0.68 eV, which is weaker than the S2-Pb2 bond. As shown in Figure 12e–h, the interaction between the O 2s and Pb 6s orbitals in TA shows a balance of bonding and anti-bonding characteristics in the range from −10 to −5 eV, indicating a minor impact on the overall adsorption process. The interaction between the O 2s and Pb 6p orbitals is mainly bonding below −5 eV, with a small anti-bonding effect between −5 and 0 eV. At the same time, the interaction between the S 3p and Pb 6s orbitals has both bonding and anti-bonding effects in the low-energy region from −10 to −5 eV. The interaction between the S 3p and Pb 6p orbitals has a relatively strong bonding effect in the range from −10 to 0 eV.

Figure 12d shows the contributions of various orbital interactions to the O1-Pb3 bond. The bond between O 2p and Pb 6p exhibits a strong negative -ICOHP value of 0.37 eV, making the greatest contribution to the bonding. The interaction between S 3s and Pb 6p comes next, with a -ICOHP value of 0.32 eV. The -ICOHP values of the S 3s-Pb 6s and S 3p-Pb 6s bonds are −0.01 eV and 0.02 eV, respectively, contributing very little to the bonding and thus can be neglected.

Based on the above analysis, it can be concluded that Z200 adsorption on the galena surface forms only a single S1–Pb1 bond with an adsorption energy of 1.00 eV. In contrast, TA adsorption on galena results in the formation of dual bonds (S2–Pb2 and O1–Pb3), yielding a combined adsorption energy of 1.76 eV, which is markedly higher than that of Z200. This enhanced interfacial binding energy demonstrates that TA exhibits stronger adsorption affinity on galena surfaces compared to Z200. It can be seen that the strength of the S1-Pb1 bond formed by the adsorption of Z200 on the surface of galena is approximately similar to that of the S2-Pb2 bond formed by the adsorption of TA on the surface of galena. The COHP analysis shows that the interactions between S 3s, S 3p, and Pb 6p contribute the most to the S1-Pb1 and S2-Pb2 bonds. In addition, after TA is adsorbed on the surface of galena, in addition to the S2-Pb2 bond, an O1-Pb3 bond can also be formed. Therefore, the adsorption energy of TA on the surface of galena is greater than that of Z200. Consequently, when Z200 is used as a collector, TA can be used as a depressant to suppress the flotation of galena.

4. Conclusions

In order to improve the utilization rate of copper resources and reduce the lead content in copper concentrates, a novel galena depressant, TA, was proposed in this study. While this study demonstrates effective separation under idealized conditions with pure minerals, further work is required to extend these findings to partially liberated minerals from the same host rock, which represents a more industrially relevant challenge. Micro-flotation tests, Fourier transform infrared spectroscopy (FTIR), X-ray photoelectron spectroscopy (XPS), and quantum chemical calculations were employed to analyze the depressant performance of TA on galena.

The results of micro-flotation tests indicated that TA significantly inhibited the flotation of galena within the pH range of 8–12 while having almost no impact on the flotation of chalcopyrite. Further FTIR analysis and molecular simulations confirmed that during the adsorption process of TA on the surface of galena, there was an electron transfer from the S and O atoms in TA to the Pb atoms on the surface of galena. This was also verified by XPS.

The COHP analysis revealed that the interactions between the 3s and 3p orbitals of the S atom in the flotation reagents (Z200 and TA) and the 6p orbital of the Pb atom on the surface of galena made significant contributions to the formation of the S-Pb bond. Additionally, when TA adsorbed on the surface of galena, it could form not only S-Pb bonds but also O-Pb bonds. This was the main reason why the adsorption ability of TA on the surface of galena was stronger than that of Z200.

While this orbital engineering framework establishes a theoretical foundation for depressant design, two critical challenges remain: (1) scale-up of reagent synthesis to meet industrial throughput requirements, and (2) water recycling compatibility in flotation circuits, where chemical interactions with recycled process water may alter reagent efficacy. Future investigations integrating pilot-scale trials and techno-economic assessments are essential to address these gaps and facilitate practical implementation.

Author Contributions: Conceptualization, H.Z., C.S. and C.L.; methodology, Z.G. and T.L.; software, R.L. and J.H.; validation, T.L.; formal analysis, Z.Z. and Z.G.; investigation, C.L., T.L. and Y.Z.; resources, C.L.; data curation, Z.Z. and Z.G.; writing—original draft preparation, H.Z.; writing—review and editing, G.W. and J.H.; visualization, Z.Z., R.L. and J.H.; supervision, Y.Z.; project administration, Y.Z.; funding acquisition, C.L. All authors have read and agreed to the published version of the manuscript.

Funding: This research was funded by The National Key Research and Development Program of China, grant 2024YFC2909602.

Data Availability Statement: Data are contained within the article.

Acknowledgments: The authors acknowledge the Beijing Super Cloud Center (BSCC) for providing HPC resources that have contributed to the research results reported within this paper. URL: http://www.blsc.cn/ (accessed on 18 February 2025). The authors extend their gratitude to Zhitong Yuan from Scientific Compass (www.shiyanjia.com) for providing invaluable assistance with the XPS analysis.

Conflicts of Interest: The authors declare no conflicts of interest.

References

1. Cui, Y.; Jiao, F.; Qin, W.; Wang, C.; Li, X. Flotation separation of sphalerite from galena using eco-friendly and efficient depressant pullulan. *Sep. Purif. Technol.* **2022**, *295*, 121013. [CrossRef]
2. Su, C.; Cai, J.; Yu, X.; Peng, R.; Zheng, Q.; Ma, Y.; Liu, D. Effect of pre-oxidation on galena in the flotation separation of chalcopyrite from galena with calcium lignosulfonate as depressant. *Miner. Eng.* **2022**, *182*, 107520. [CrossRef]
3. Lai, H.; Liu, Q.; Deng, J.; Wen, S.; Liu, Z. Surface chemistry study of Cu-Pb sulfide ore using ToF-SIMS and multivariate analysis. *Appl. Surf. Sci.* **2020**, *518*, 146270. [CrossRef]
4. Lai, H.; Liu, Q.; Deng, J.; Wen, S. Using ToF-SIMS to study metal ions transfer between chalcopyrite and galena during grinding. *Adv. Powder Technol.* **2020**, *31*, 2650–2657. [CrossRef]
5. Cai, J.; Jia, X.; Ma, Y.; Ibrahim, A.M.; Su, C.; Yu, X.; Liu, D. Effect of Pre-Oxidation on Copper-Lead Bulk Concentrate Flotation Separation with Sodium Polyacrylate as Galena Depressant. *Sep. Purif. Technol.* **2022**, *302*, 122276. [CrossRef]
6. Yang, X.; Lai, H.; Wei, X.; Shen, P.; Wang, Y.; Li, M.; Cai, J.; Liu, D. Adsorption Mechanism of $Na_2S_2O_3$ and $FeSO_4$ as a Combined Depressant for Galena in Chalcopyrite-Galena Flotation Separation. *Colloids Surf. A Physicochem. Eng. Asp.* **2025**, *706*, 135799. [CrossRef]
7. Wills, B. The Separation by Flotation of Copper–Lead–Zinc Sulphides. *Min. Mag.* **1984**, *150*, 36–41.
8. Ralston, J. The chemistry of galena flotation: Principles & practice. *Miner. Eng.* **1994**, *7*, 715–735.
9. Okada, S.; Majima, H. Depressive action of chromate and dichromate salts on galena. *Can. Metall. Q.* **1971**, *10*, 189–195. [CrossRef]
10. Zhang, X.; Lu, L.; Zeng, H.; Hu, Z.; Zhu, Y.; Han, L. A macromolecular depressant for galena and its flotation behavior in the separation from molybdenite. *Miner. Eng.* **2020**, *157*, 106576. [CrossRef]
11. Zhang, X.; Qian, Z.; Zheng, G.; Zhu, Y.; Wu, W. The design of a macromolecular depressant for galena based on DFT studies and its application. *Miner. Eng.* **2017**, *112*, 50–56. [CrossRef]
12. Pugh, R. Macromolecular organic depressants in sulphide flotation—A review, 1. Principles, types and applications. *Int. J. Miner. Process.* **1989**, *25*, 101–130. [CrossRef]
13. Cao, J.; Liao, R.; Wu, D.; Zuo, Q.; Liu, J.; Wen, S. Utilizing Phosphonyl Carboxylic Acid Copolymer as an Efficient Depressant for Flotation Separation of Chalcopyrite from Galena: Experimental and DFT Study. *Sep. Purif. Technol.* **2024**, *348*, 127725. [CrossRef]
14. Liu, D.; Zhang, G.; Chen, Y. Investigations on the selective depression of fenugreek gum towards galena and its role in chalcopyrite-galena flotation separation. *Miner. Eng.* **2021**, *166*, 106886. [CrossRef]
15. Miao, Y.; Wen, S.; Shen, Z.; Feng, Q.; Zhang, Q. Flotation separation of chalcopyrite from galena using locust bean gum as a selective and eco-friendly depressant. *Sep. Purif. Technol.* **2022**, *283*, 120173. [CrossRef]
16. Qin, W.; Wei, Q.; Jiao, F.; Li, N.; Wang, P.; Ke, L. Effect of sodium pyrophosphate on the flotation separation of chalcopyrite from galena. *Int. J. Min. Sci. Technol.* **2012**, *22*, 345–349. [CrossRef]
17. Huang, P.; Cao, M.; Liu, Q. Using chitosan as a selective depressant in the differential flotation of Cu–Pb sulfides. *Int. J. Miner. Process.* **2012**, *106*, 8–15. [CrossRef]
18. Pugh, R. Macromolecular organic depressants in sulphide flotation—A review, 2. Theoretical analysis of the forces involved in the depressant action. *Int. J. Miner. Process.* **1989**, *25*, 131–146. [CrossRef]
19. Pan, Z.; Cai, C.; Liang, G.; Jiao, F.; Yang, C.; Wei, Q.; Qin, W. Pyrolysis Liquid of Pine Wood: A Novel Efficient Depressant of Pyrite in Galena Flotation. *Miner. Eng.* **2024**, *210*, 108642. [CrossRef]

20. Wang, Z.; Han, G.; Feng, Q. Selective Flotation of Galena and Sphalerite Using N-(Phosphonomethyl) Iminodiacetic Acid as an Eco-Friendly Depressant. *Green Smart Min. Eng.* **2024**, *1*, 96–103. [CrossRef]
21. Zhang, H.; Cao, Q.; Yan, Y.; Zou, H.; Huang, X.; Liu, D. Flotation Separation of Galena and Chalcopyrite by Using Hydroxyl Radicals from an Fe^{2+}/NaClO System as Depressants. *Appl. Surf. Sci.* **2025**, *686*, 162128. [CrossRef]
22. Hu, Y.; Zhao, Z.; Lu, L.; Zhu, H.; Xiong, W.; Zhu, Y.; Yang, B. Investigation on a Novel Galena Depressant in the Flotation Separation from Molybdenite. *Minerals* **2021**, *11*, 12. [CrossRef]
23. Liu, C.; Wu, G.; Yanhong, M.; Lu, T.; Liu, H.; Zhao, Z.; Chen, L. Application and Quantum Chemical Analysis of Novel Sulfur-Containing Heterocyclic Inhibitors in Separation of Molybdenite and Galena. *Conserv. Util. Miner. Resour.* **2022**, *42*, 75–81.
24. Zhu, H.; Yang, B.; Martin, R.; Zhang, H.; He, D.; Luo, H. Flotation Separation of Galena from Sphalerite Using Hyaluronic Acid (HA) as an Environmental-Friendly Sphalerite Depressant. *Miner. Eng.* **2022**, *187*, 107771. [CrossRef]
25. Zhai, Q.; Dong, W.; Liu, R.; Xie, Z.; Cao, Z.; Sun, W. Green Separation of Galena from Molybdenite by Flotation Using DL-Dithiothreitol as a Depressant. *Sep. Purif. Technol.* **2024**, *347*, 127676. [CrossRef]
26. Pearse, M.J. Historical Use and Future Development of Chemicals for Solid–Liquid Separation in the Mineral Processing Industry. *Miner. Eng.* **2003**, *16*, 103–108. [CrossRef]
27. Moudgil, B.M. Effect of Polyacrylamide and Polyethylene Oxide Polymers on Coal Flotation. *Colloids Surf.* **1983**, *8*, 225–228. [CrossRef]
28. Boulton, A.; Fornasiero, D.; Ralston, J. Selective Depression of Pyrite with Polyacrylamide Polymers. *Int. J. Miner. Process.* **2001**, *61*, 13–22. [CrossRef]
29. Kresse, G.; Furthmüller, J. Efficient Iterative Schemes for Ab Initio Total-Energy Calculations Using a Plane-Wave Basis Set. *Phys. Rev. B* **1996**, *54*, 11169–11186. [CrossRef]
30. Blöchl, P.E. Projector Augmented-Wave Method. *Phys. Rev. B* **1994**, *50*, 17953–17979. [CrossRef]
31. Gillan, M.J.; Alfe, D.; Michaelides, A. Perspective: How Good Is DFT for Water? *J. Chem. Phys.* **2016**, *144*, 130901. [CrossRef] [PubMed]
32. Smith, D.G.A.; Burns, L.A.; Patkowski, K.; Sherrill, C.D. Revised Damping Parameters for the D3 Dispersion Correction to Density Functional Theory. *J. Phys. Chem. Lett.* **2016**, *7*, 2197–2203. [CrossRef]
33. Lu, T.; Chen, F. Revealing the Nature of Intermolecular Interaction and Configurational Preference of the Nonpolar Molecular Dimers $(H_2)_2$, $(N_2)_2$, and $(H_2)(N_2)$. *J. Mol. Model.* **2013**, *19*, 5387–5395. [CrossRef]
34. Schimka, L.; Harl, J.; Stroppa, A.; Grüneis, A.; Marsman, M.; Mittendorfer, F.; Kresse, G. Accurate Surface and Adsorption Energies from Many-Body Perturbation Theory. *Nat. Mater.* **2010**, *9*, 741–744. [CrossRef]
35. Momma, K.; Izumi, F. VESTA 3 for Three-Dimensional Visualization of Crystal, Volumetric and Morphology Data. *J. Appl. Crystallogr.* **2011**, *44*, 1272–1276. [CrossRef]
36. Noda, Y.; Masumoto, K.; Ohba, S.; Saito, Y.; Toriumi, K.; Iwata, Y.; Shibuya, I. Temperature Dependence of Atomic Thermal Parameters of Lead Chalcogenides, PbS, PbSe and PbTe. *Cryst. Struct. Commun.* **1987**, *43*, 1443–1445. [CrossRef]
37. Naumov, P.; Makreski, P.; Petruševski, G.; Runčevski, T.; Jovanovski, G. Visualization of a Discrete Solid-State Process with Steady-State X-Ray Diffraction: Observation of Hopping of Sulfur Atoms in Single Crystals of Realgar. *J. Am. Chem. Soc.* **2010**, *132*, 11398–11401. [CrossRef] [PubMed]
38. Ke, B.; Li, Y.; Chen, J.; Zhao, C.; Chen, Y. DFT Study on the Galvanic Interaction between Pyrite (100) and Galena (100) Surfaces. *Appl. Surf. Sci.* **2016**, *367*, 270–276. [CrossRef]
39. Ke, B.; Chen, J.; Cheng, W. Galvanic Interaction between Different Grinding Media and Galena (1 0 0) Surface and Its Influence on Galena Flotation Behavior: A DFT Study. *Appl. Surf. Sci.* **2021**, *570*, 151379. [CrossRef]
40. Feng, B.; Zhang, W.; Guo, Y.; Wang, T.; Luo, G.; Wang, H.; He, G. The flotation separation of galena and pyrite using serpentine as depressant. *Powder Technol.* **2019**, *342*, 486–490. [CrossRef]
41. Mikhlin, Y.L.; Karacharov, A.A.; Likhatski, M.N. Effect of adsorption of butyl xanthate on galena, PbS, and HOPG surfaces as studied by atomic force microscopy and spectroscopy and XPS. *Int. J. Miner. Process.* **2015**, *144*, 81–89. [CrossRef]
42. Zhao, Z.; Gao, Z.; Sun, C.; Zhang, Y.; Liu, C.; Lu, T.; Hu, J. Novel Depressant Based on Hybridization Theory for the Separation of Galena and Realgar: Experimental Study and Adsorption Mechanism. *Minerals* **2025**, *15*, 200. [CrossRef]
43. Gao, Z.; Liu, C.; Lu, T.; Zhao, Z.; Wu, G.; Zhu, Y. Understanding the Adsorption Mechanism of BTPA, DEPA, and DPPA in the Separation of Malachite from Calcite and Quartz: DFT and Experimental Studies. *Minerals* **2024**, *14*, 692. [CrossRef]
44. Cui, Q.; Liu, C.; Gao, Z.; Lu, T.; Zhao, Z.; Zhu, Y.; Wu, G. First Principle Study of the Relationship between Electronic Properties and Adsorption Energy: Xanthate Adsorption on Pyrite and Arsenopyrite. *Minerals* **2024**, *14*, 749. [CrossRef]
45. Miao, Y.; He, J.; Zhu, X.; Zhu, G.; Cao, S.; Fan, G.; Li, G.; Cao, Y. Hardness of Surface Hydroxyls and Its Pivotal Role in the Flotation of Cassiterite from Quartz via Lead Ions Activation. *Sep. Purif. Technol.* **2024**, *347*, 127565. [CrossRef]

46. Liu, J.; Zhu, J.; Xu, H.; Cheng, D. Rational Design of Heteroatom-Doped Fe–N–C Single-Atom Catalysts for Oxygen Reduction Reaction via Simple Descriptor. *ACS Catal.* **2024**, *14*, 6952–6964. [CrossRef]

MDPI AG
Grosspeteranlage 5
4052 Basel
Switzerland
Tel.: +41 61 683 77 34

Minerals Editorial Office
E-mail: minerals@mdpi.com
www.mdpi.com/journal/minerals

www.ingramcontent.com/pod-product-compliance
Lightning Source LLC
Chambersburg PA
CBHW061609210926
51850CB00009B/2262